Solar Cooling

Cooling buildings is a major global energy consumer and the energy requirement is growing year by year. This guide to solar cooling technology explains all you need to know about how solar energy can be converted into cooling energy. It outlines the difference between heat-driven and photovoltaic-driven systems and gives examples of both, making clear in what situations solar cooling technology makes sense. It includes chapters on:

- solar thermal collectors
- solar cooling technologies
- cold distribution
- storage components
- designing and sizing
- installation, operation and maintenance
- economic feasibility
- potential markets
- case studies.

Solar Cooling is for engineers, architects, consultancies, solar thermal technology companies, students and anyone who is interested in getting involved with this technology.

Paul Kohlenbach is a Professor at the Beuth University of Applied Sciences Berlin, Germany. His area of expertise is renewable energies, including but not limited to solar thermal process heat, solar cooling, concentrating solar power (CSP), trigeneration, façade integrated solar systems, solar hot water systems, hydrogen technology and fuel cells. He is also the founder and owner of Solem Consulting, an international consortium of leading experts in solar technologies which was founded in 2008. Over 14 years of professional experience with solar thermal systems, especially in engineering design, planning and control of such, make him an internationally renowned solar expert.

Uli Jakob is the General Manager of Solem Consulting in Europe and an internationally renowned solar cooling expert. His areas of expertise are building physics, renewable energies and climate engineering, with a special focus on innovative energy concepts and energy-efficient buildings. He is also founder and owner of dr jakob energy research, a consultancy for research, product development and marketing with a focus on renewables, energy-efficient buildings and industry. Dr Jakob has over 14 years of professional experience, especially with the product development and commercialisation of solar/thermal cooling kits. He is also a lecturer at the University of Applied Sciences Stuttgart, Germany and Hamburger Fern-Hochschule, Germany.

Earthscan Expert Series
Series editor: Frank Jackson

Solar:

Grid-Connected Solar Electric Systems
Geoff Stapleton and Susan Neill

Pico-solar Electric Systems
John Keane

Solar Cooling
Paul Kohlenbach and Uli Jakob

Solar Domestic Water Heating
Chris Laughton

Solar Technology
David Thorpe

Stand-alone Solar Electric Systems
Mark Hankins

Home Refurbishment:

Sustainable Home Refurbishment
David Thorpe

Wood Heating:

Wood Pellet Heating Systems
Dilwyn Jenkins

Renewable Power:

Renewable Energy Systems
Dilwyn Jenkins

Energy Management:

Energy Management in Buildings
David Thorpe

Energy Management in Industry
David Thorpe

Solar Cooling

The Earthscan Expert Guide to Solar Cooling Systems

Paul Kohlenbach and Uli Jakob

Routledge
Taylor & Francis Group
LONDON AND NEW YORK

earthscan
from Routledge

First edition published 2014 by Routledge
2 Park Square, Milton Park, Abingdon, Oxon OX14 4RN
52 Vanderbilt Avenue, New York, NY 10017

Routledge is an imprint of the Taylor & Francis Group, an informa business

British Library Cataloguing in Publication Data
A catalogue record for this book is available from the British Library

Library of Congress Cataloging in Publication Data
Kohlenbach, Paul.
Solar cooling : the Earthscan expert guide to solar cooling systems /
Paul Kohlenbach and Uli Jakob. -- First edition.
pages cm -- (Earthscan expert)
Includes bibliographical references and index.
1. Solar air conditioning. I. Jakob, Uli. II. Earthscan. III. Title.
TH7687.9K64 2014
697.9'3--dc23
2013046461

ISBN 13: 978-0-415-63975-0 (hbk)

Typeset in Sabon by Fakenham Prepress Solutions, Fakenham,
Norfolk NR21 8NN

Contents

List of figures

List of tables

Acknowledgements

This book has required a lot of time and effort. Therefore, we would like to express our heartfelt thanks to our families, who have always supported us in the process of writing this book.

We would like to acknowledge the contributions of various people to this book. We are very grateful for contributions from:

- Dr Mike Dennis, Australian National University (Chapter 5.2.3)
- Dr Daniel Mugnier, Tecsol (Chapter 3.5)
- Jeremy Osborne, Solem Consulting (Chapter 10.8)

We are grateful to Frank Jackson for his thorough editing. He provided valuable advice on how to transfer the technical content to the target audiences. We also thank all contributors of graphic material and pictures for their support.

1 Introduction

This book aims to give an overview on the technological and economical aspects of solar cooling systems for buildings, concentrating on larger solar thermal cooling systems. The title term 'solar cooling' refers to two categories of systems:

1. Systems providing air-conditioning, e.g. for buildings, usually in the form of either chilled water or air.
2. Systems providing cold, e.g. for industrial processes, usually in the form of chilled water.

The book's focus is on the provision of air-conditioning for buildings, but systems providing cold for industrial processes are also discussed.

1.1 Active and passive cooling systems

There are two kinds of cooling systems: active and passive. Active systems are classified as those where energy, typically electricity and heat, is used to drive a thermal conversion process that provides some form of cold. Passive systems are systems without any external energy input. Large solar-powered active systems – systems providing cooling for entire buildings, not small solar-assisted units for single rooms – are the primary focus of this book, though passive systems will also be briefly discussed.

Worldwide, the energy consumption required for active cooling and air-conditioning is rising rapidly. Standard electrically driven compressor chillers and air-conditioners (typically reverse-cycle split units) have seen a continuous global increase in sales. In particular, the sales figures for split units with a cooling capacity range up to $5\,kW_r$ have risen rapidly in recent years. The Japan Refrigeration and Air Conditioning Industry Association (JRAIA) has estimated worldwide sales of these systems to be 93.8 million units in 2012 [1]. To put this in perspective, the number of cars sold worldwide in 2012 was 81 million units [4].

Air-conditioners and chillers draw their maximum energy consumption during their peak-load periods during the summer. In the last few years this has regularly led to grids working to maximum capacity and blackouts in, for example, southern Europe and other regions of the world. Conventional small air-conditioners (also known as split units) have high energy consumption and use non-environmentally friendly refrigerants. The refrigerants currently used in split units no longer have an ozone depletion potential (ODP), but they have a considerable global warming potential (GWP) because of refrigerant leakages in the range of 5–15 per cent per year.

Passive systems – systems without any external energy input – include, for example, solar shading devices, semi-transparent glass covers or solar-driven ventilation effects. Passive cooling depends on building construction and thus needs to be considered when designing a new building. One passive cooling option is to design a heavy-walled building (e.g. with concrete walls and ceilings), which reduces the influence of solar radiation and internal loads on the cooling requirement. The heat-storage capacity of the building material provides thermal insulation which keeps heat out of the building. A second option that can be used in new light-walled buildings are phase-change materials (PCMs) which can be integrated into wall plaster or suspended ceilings. The building heat will then be taken up in the PCM during the day, thus cooling the building, and be discharged to the building during night-time to be removed by night ventilation. The PCM option can be applied to new and existing buildings without major rebuilding. Other options are passive cooling systems that prevent external heat input into the building. These include, for example, solar shading devices, semi-transparent glass covers and natural ventilation effects. Passive cooling measures reduce the building's cooling load. The remaining cooling load is now low enough to be more efficiently covered using a smaller active cooling system. In other words, every unit of heat that is prevented from entering the building through passive cooling measures will save electricity used by the active cooling system.

1.2 Solar cooling

Active solar cooling systems are an attractive alternative to conventional electrically driven air-conditioning. They can be used in virtually all applications where conventional air-conditioning or cooling is used. There are two ways to harness the solar energy to drive active solar cooling systems:

1. solar thermal collectors, converting sun rays into heat; or
2. solar photovoltaic panels, converting sun rays into electrical power.

The difference between the two above possibilities is quite distinct. This book discusses both options in Chapter 3. However, the focus of this book is on solar thermal systems (option 1) as they currently have a higher market share. These combine thermally driven sorption chillers with solar thermal collectors. Sorption chillers use environmentally friendly refrigerants (water or ammonia) and have only very low electricity demand. Therefore the operating costs of these chillers are very low and the CO_2 balance compared to conventional systems is considerably better. In the case of active solar cooling the main advantage is the coincidence of solar irradiation and cooling demand, which matches very well in sunny and hot climates all over the world.

At the time of writing about 1,000 large solar cooling systems are installed worldwide and the market has grown in the last eight years between 40 and 70 per cent per year. The overall number of systems installed to date indicates that solar cooling is still a niche market, but one which is developing. Since 2007, a system cost reduction of about 50 per cent was realized because of further standardization of pre-grouped components, so-called solar cooling kits. Solar cooling is especially appealing if the solar thermal system is also used for other applications such as heating, hot water, etc. Maximum solar system operation time and a low-cost source of driving heat for sorption chillers are key factors in making solar thermal cooling systems economically feasible.

1.2.1 Pre-conditions for solar cooling

All solar cooling systems require three main boundary conditions in order to work well. First, a good solar resource needs to be available – the more sun can be utilised the better. This depends on the geographic location of the building. Second, the climate zone needs to be taken into consideration. Seasonal and daily variations of local climate determine the heating and cooling loads of a building. Third, energy costs should be considered – electricity and fossil fuel costs determine the economics of a solar cooling system. A fourth, less important, condition is the simultaneity of solar radiation and cooling requirement of a building. In other words, the building should ideally have its heating/cooling loads when the sun is shining during the day, and not early in the morning or at night. These boundary conditions are discussed in more detail in this book.

1.2.2 A brief history of solar cooling

The world's first solar cooling system was operated in Paris, France, during the 1878 world exhibition [2]. The system consisted of an ammonia–water ABsorption chiller and a parabolic reflector, and produced ice blocks. It was purely a demonstration system. The first commercial solar cooling systems for air-conditioning were developed in Europe and the USA 100 years later, in the 1970s, by companies such as Dornier-Prinz Solartechnik (Germany), Arkla Industries (USA, today Robur/Italy) and Carrier (USA). Several demonstration projects were built, motivated by high fossil fuel cost during the global oil crisis of the late 1970s. After oil prices decreased again, the lack of demand on the market for solar cooling resulted in the production of solar cooling systems ceasing in the late 1980s [3].

In the 1990s companies like Yazaki (Japan) and Thermolux (Germany) manufactured and installed a few custom-made solar cooling systems. After 2000, the first solar cooling system suppliers started offering solar cooling systems. Contrary to earlier attempts, this was done with a focus on the whole system, not just the individual components. Typically, these were solar thermal collector suppliers looking to increase their sales volume by including a cooling option with their products. Suppliers included Citrin Solar, Conergy and SolarNext (all from Germany) and SOLID and Sol-ution (both from Austria), among others. The demand led to a number of manufacturers of small-scale

ABsorption or ADsorption chillers entering the market. Today, companies like SorTech AG, InvenSor GmbH, EAW (all from Germany), Pink (Austria), SolabCool (Netherlands), Yazaki (Japan) and Thermax (India) provide small capacity chillers with a cooling capacity between $5\,kW_r$ and $35\,kW_r$. As we write, the solar cooling market is far from being established; however, rising energy costs and increasing public awareness of environmental issues are likely to change this.

2
Solar thermal collectors

In this chapter we discuss the workings and general composition of solar thermal collectors. The chapter covers all solar thermal collectors suitable for solar cooling systems. The physics behind the technology is also discussed, especially the collector efficiencies.

Essential terms for solar thermal collectors

Absorptivity – the fraction of radiation absorbed at a specific wavelength

Absorber area – the area of a solar thermal collector where radiation energy is converted into thermal energy

Aperture area – the area of a solar thermal collector that collects the sun's radiation energy (typically larger than the absorber area)

Concentration ratio – the ratio of aperture to absorber area of a concentrating solar collector

Diffuse radiation – solar radiation that interacts with the atmosphere before reaching the Earth's surface. Diffuse radiation is non-directional. It is caused by scattering of light at molecules in the atmosphere, reflection and absorption/re-emission

Direct radiation – solar radiation that does not interact with the atmosphere before reaching the Earth's surface. Direct radiation is directional and comes straight from the sun

DNI – direct normal irradiation; the direct radiation falling orthogonally onto a collector surface

Emissivity – the relative ability of a material surface to emit energy by radiation. It is expressed as the ratio of energy radiated by a material at a given temperature to energy radiated by a black body at the same temperature

Gross area – the total area a solar thermal collector requires for complete installation on a roof or the ground. Gross area includes space required for access and peripherals

Heat convection – a heat transfer process based on the movement of liquids or gases

Incidence angle – the angle between a light ray and the collector glass cover

IAM – incident angle modifier, which accounts for the influence of non-perpendicular incident radiation at incidence angle Θ in relation to normal incidence radiation ($\Theta = 0$)

Optical collector efficiency – the maximum efficiency of a solar collector. The optical efficiency denotes the optical losses of a collector, consisting of reflection and absorption in the glass cover as well as reflection from the absorber sheet. It is the point of the efficiency curve where the mean collector temperature equals the environment temperature

Reflectivity – the fraction of radiation that is reflected at a surface

Thermal conductivity – the ability of a substance to conduct heat

2.1 Solar thermal collectors

Solar thermal collectors come in various configurations but the basic principle behind them is the same for all: the conversion of radiation energy from sunlight into thermal energy, i.e. heat. This heat can then be used to heat water or air. One of the simplest ways of heating water with the sun is to expose a black hose to strong sunshine. Water flowing through the hose heats up. Also, camping showers are available which are simple black plastic bags filled with water. Placed in the sun the water heats up and can be used for showers.

This energy conversion from sunlight to heat is based on the absorption of light by a surface. How much energy is absorbed depends on the spectral composition of the light as well as the material of the surface. Not all wavelengths in the sunlight spectrum are absorbed equally. Most sunlight energy that is converted in solar thermal collectors (approx. 50 per cent) comes from the visible part of the sunlight spectrum (which ranges across approx. 400–780 nanometres of wavelength). Only a small fraction of sunlight energy (approx. 10 per cent) is contained in ultraviolet light (200–400 nanometre range). Infrared light and longer radio waves (>780 nanometres) contain approx. 40 per cent of the total energy contained in sunlight. A black surface enhances the absorptivity of the energy conversion.

If the energy-absorbing surface is in contact with either a still reservoir of water (e.g. the camping shower mentioned above) or a water flow (e.g. a garden hose or a solar thermal collector) then heat from the surface is transferred to the water. This is the basic principle of a solar thermal collector (see Figure 2.1).

All commercially available solar thermal collectors suitable for solar cooling are explained in the following chapters.

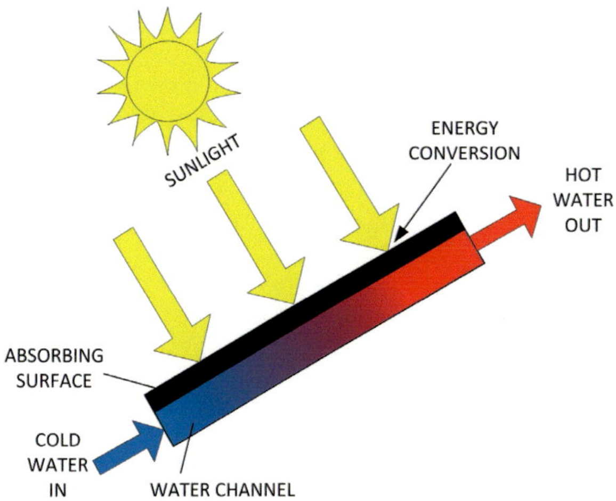

Figure 2.1 Basic principle of a solar thermal collector. Sunlight (radiation energy) is converted into heat (thermal energy). The heat is transferred to a water flow.

2.1.1 Flat plate collectors

Flat plate collectors usually consist of the following main components:

- a copper or aluminium absorber sheet, selectively coated
- copper or aluminium pipes welded to the back of the absorber sheet
- a glass cover
- an aluminium frame with a sheet metal backside
- thermal insulation.

The general layout is shown in Figure 2.2.

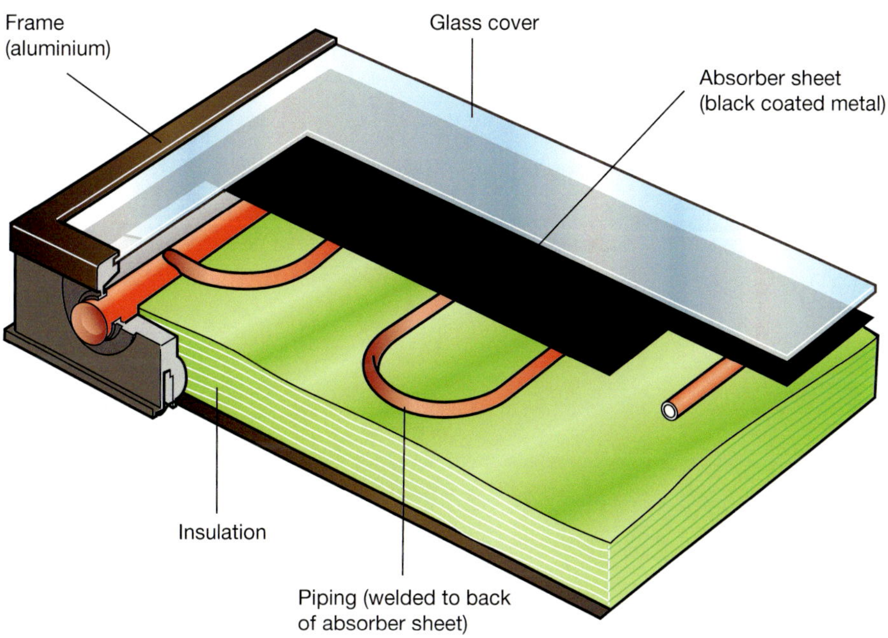

Figure 2.2 General layout of a flat plate collector.

The absorber collects the sun's radiation and transforms it into heat. Heat transfer from the absorber to the piping results in a temperature increase of the fluid flowing through the collector piping. The most common fluid is water, which in frost-prone areas is mixed with glycol to avoid freezing. The underside of the collector is insulated to minimise thermal loss. The glass cover reduces heat loss due to convection (e.g. caused by wind). The absorber sheet is usually coated with a layer of spectrally selective material. This so-called selective coating improves the absorptivity of the sunlight's energy in the visible wavelengths. It also reduces the emissivity of heat radiation in the non-visible infrared wavelengths, thus increasing the amount of heat transferred to the fluid. Globally, flat plate collectors are the most common solar collectors for domestic hot water and space heating. An exception is China, where evacuated tube collectors are mostly used for this purpose. The advantages and disadvantages of flat plate collectors for solar cooling systems are shown in Table 2.1.

Flat plate collectors are limited in the temperature level they can provide compared to evacuated tubes and concentrating collectors (see Figure 2.20). They are therefore suitable only for solar cooling applications with driving temperatures below 100 °C (212 °F). The lower collector efficiency at temperatures greater than 100 °C (212 °F) results in an increased collector area to provide the same thermal power.

Table 2.1 Advantages and disadvantages of flat plate collectors for solar cooling systems

Advantages	Disadvantages
• simple and proven technology • readily available • no moving parts • good cost–performance ratio • easy installation on roof and ground surface	• temperature limit of approx. 100 °C (212 °F) • lower efficiency

Figure 2.3 Example of flat plate collectors on a commercial building roof in Berlin, Germany. Note the dual return pipe on the right-hand side of the picture for balanced flow (Tichelmann piping; see Chapter 6.3.5).

Source: Solem Consulting

2.1.2 Evacuated tube collectors

Evacuated tube collectors are an improved version of flat plate collectors with regard to thermal losses. Here, the absorber sheet is encased in a glass tube from which the air has been evacuated. The vacuum in the glass tube results in less heat being lost to the environment and thus a greater collector efficiency. Evacuated tube collectors are commercially available in two different models:

1. direct flow collectors
2. heat pipe collectors.

The direct flow model uses a flat rectangular or a cylindrically shaped absorber inside an evacuated glass tube. The glass tube can be single- or double-walled. In single-walled tubes, the vacuum takes up the entire inner volume of the tube. In double-walled tubes the vacuum is only present in the gap between inner and outer glass wall. Figures 2.4 and 2.5 show the cross-sections of both direct flow collector models.

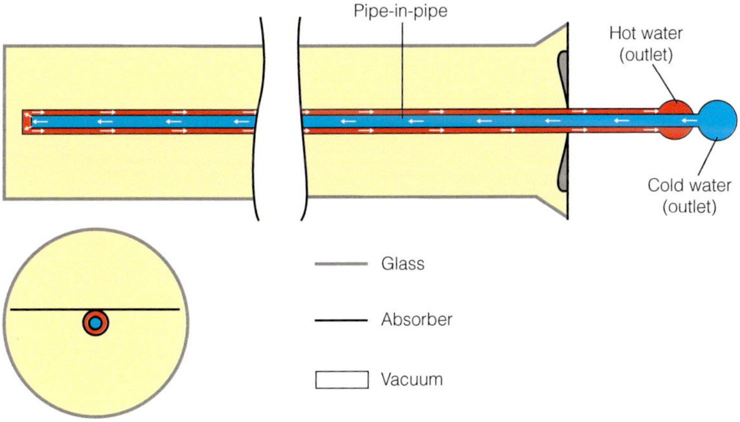

Figure 2.4 Direct flow evacuated tube with flat absorber and single-wall glass tube.

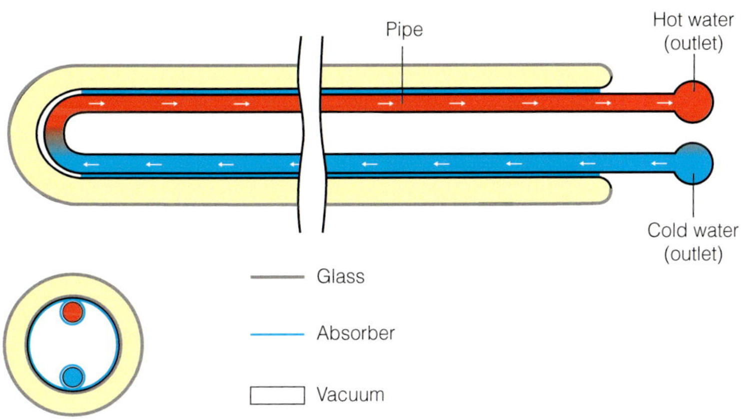

Figure 2.5 Direct flow evacuated tube with cylindrical absorber and double-wall glass tube (Sydney tube).

In direct flow models, the heat transfer fluid that is used in the system flows through the evacuated tube and is in direct contact with the absorber sheet. This is done using either a pipe-in-pipe arrangement (see Figure 2.4) or a U-shaped single pipe (see Figure 2.5). The absorber is usually spectrally selective coated. The vacuum is designed to last over the lifetime of the collector; however, atmospheric hydrogen still enters the tube in small quantities. A chemical material is used to absorb the hydrogen inside the tube and maintain the vacuum. This material is called a 'getter'. It is placed in the tube during manufacturing and remains there untouched until the end of the lifetime.

Heat pipe evacuated tube collectors have no direct heat transfer fluid flow through the evacuated tube. Instead, a secondary fluid is used to transfer the heat from the spectrally selective coated absorber to the system fluid. This is done using a heat exchanger (head) at the top of the evacuated tube (see Figures 2.7 and 2.8). The secondary fluid inside the collector pipe has a lower boiling point than water. It evaporates through heat transfer from the absorber. The vapour rises to the top of the heat pipe, where the heat pipe head is cooled

Figure 2.6 View of a direct flow, flat absorber type evacuated tube collector field on a commercial building roof in Berlin, Germany.

Source: Solem Consulting

Table 2.2 Advantages and disadvantages of evacuated tube collectors for solar cooling systems

Advantages	Disadvantages
• proven technology	• high stagnation temperature up to 200 °C (392 °F)
• readily available	• higher cost
• no moving parts	• heat pipe models require mounting at an angle
• higher temperatures possible	
• higher efficiencies	
• the individual tubes are lightweight and easy to install	
• a damaged tube can be replaced easily	

by the system heat transfer fluid passing through the collector header. The vapour condenses at the top of the heat pipe, releasing the heat previously taken up, and flows back to the bottom of the pipe where it evaporates again. The heat pipe collector has to be mounted at an angle greater than 20° and lower than 75° to the horizontal to allow the vapour to rise and the liquid to flow back.

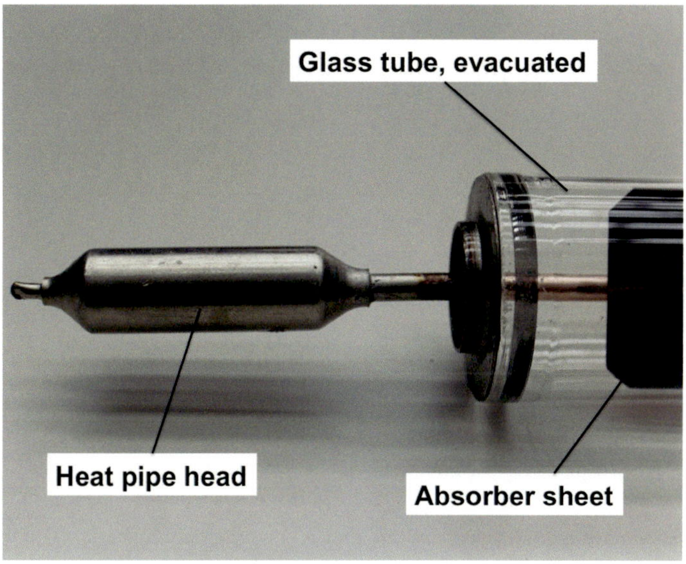

Figure 2.7 View of a heat pipe evacuated tube collector.

Source: Solem Consulting

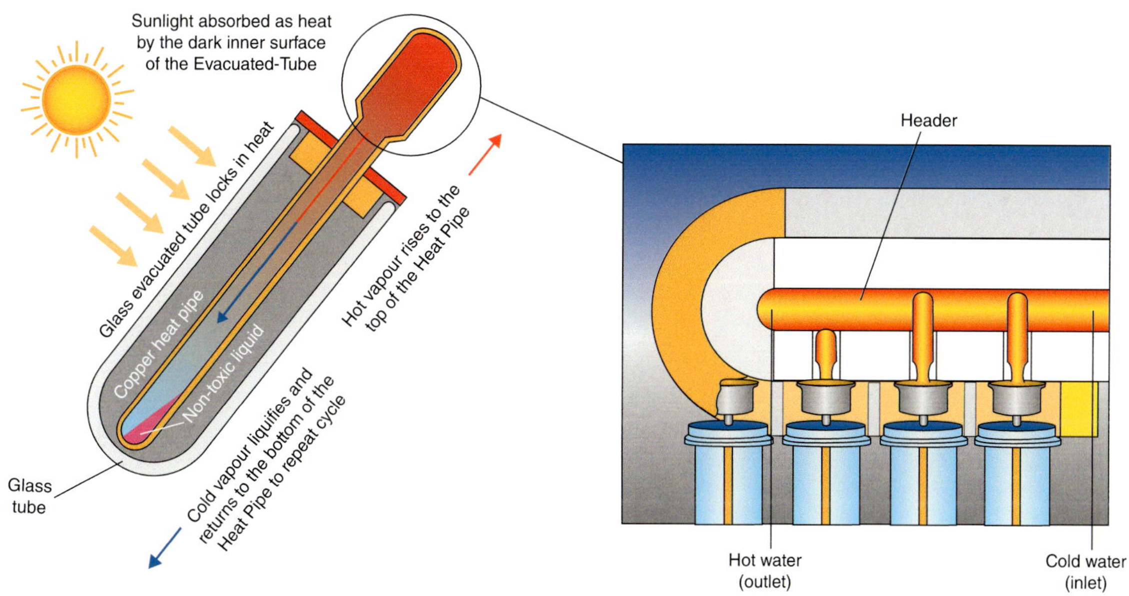

Figure 2.8 Heat pipe evacuated tube collector with cylindrical absorber and double-walled glass tube.

2.1.3 Air collectors

Air collectors are available in two different technologies. The most common air collectors have a similar construction to flat plate collectors, except the heat transfer fluid is air, not water. The absorber sheet is connected to a series of channels with air flowing through, or the absorber itself is composed of individual channels (see Figure 2.9). These air collectors typically have a glass cover, a back sheet and thermal insulation just like flat plate collectors. An air filter is usually integrated into the collector. Air collector modules also are available with built-in fans, and some models even offer a small photovoltaic module integrated in the collector to power the fan. Typical recommended air flow rates are 20–100 $m^3/(h \cdot m^2)$.

Another alternative for solar air collectors is an evacuated tube collector modified for the use of air instead of water as heat transfer fluid (see Figure 2.10). This collector can reach air temperatures up to 120 °C (248 °F) at higher efficiencies than the flat version.

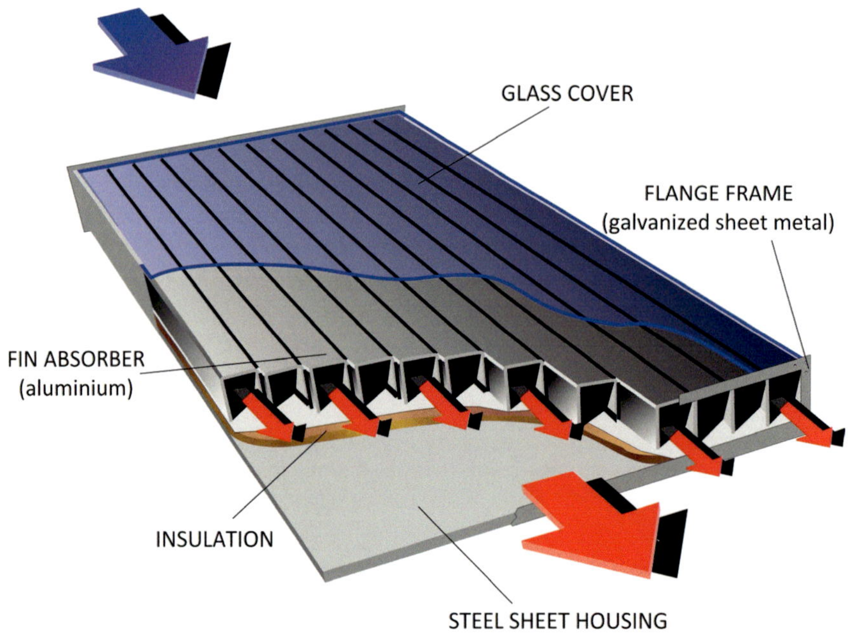

Figure 2.9 Schematic of an air collector.

Source: Grammer Solar GmbH, Germany

Table 2.3 Advantages and disadvantages of air collectors for solar cooling systems

Advantages	Disadvantages
• simple construction • easy installation • no stagnation problems during standstill	• domestic hot water generation only possible with additional heat exchanger • air ducts require more space than water piping in a building • air fan power is greater than pump power for fluid transport

Figure 2.10 View of an evacuated tube air collector.

Source: Kollektorfabrik, Germany

A disadvantage of solar air collectors is that the hot air cannot directly be used to generate domestic hot water. Manufacturers of solar air collectors offer separate heat exchangers for this purpose. The lower heat capacity of air (compared to water) results in larger air ducts required for solar air collectors.

2.1.4 Concentrating collectors

Three types of concentrating collectors are currently being used for solar air-conditioning: compound parabolic concentrating collectors (CPC), parabolic trough and linear Fresnel. They differ in the concentration ratio, the supply temperature and the tracking requirements (see Table 2.8 for details).

Compound parabolic concentrating collectors (CPC)

Compound parabolic concentrating collectors typically use a Sydney-type evacuated tube (see Chapter 2.1.2) with parabolic reflectors mounted below. Figures 2.11 and 2.12 show the tube and reflector arrangement. These reflectors are typically made of aluminium coated with a reflective layer. Reflectivity is typically around 85–90 per cent, concentration ratios are between 1 and 2.

The reflector directs solar radiation onto the sides and underside of the tube. The absorber tube therefore receives radiation directly from the top and, via reflection, on the bottom and sides. Compared to a standard evacuated tube collector, the CPC tube receives more radiation onto the absorber surface, and therefore supply temperature can be higher. However, the gross area required on the roof is also greater due to the extra space required for the reflector between the tubes. CPC collectors require regular maintenance, especially cleaning of the reflectors. The reflectors can become dirty due to dust or organic matter, which reduces the reflectivity dramatically. Then, the benefit of extra radiation on the underside of the tube is lost and the CPC operates like a standard evacuated tube collector. Some CPC models have a transparent cover over the tube and reflector to avoid this.

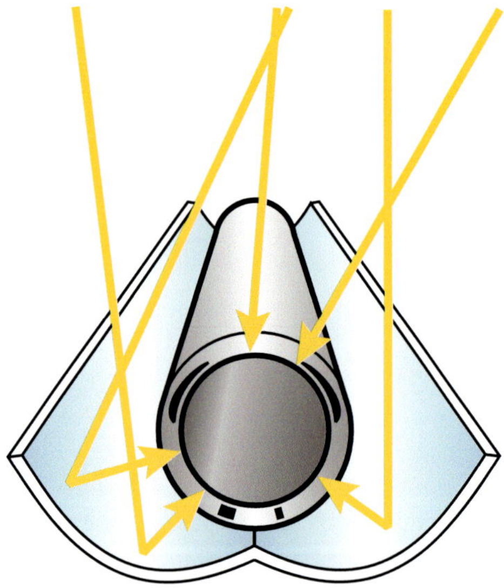

Figure 2.11 Schematic of a CPC collector. The parabolic reflector directs the solar radiation to the largest possible surface area of the tube.

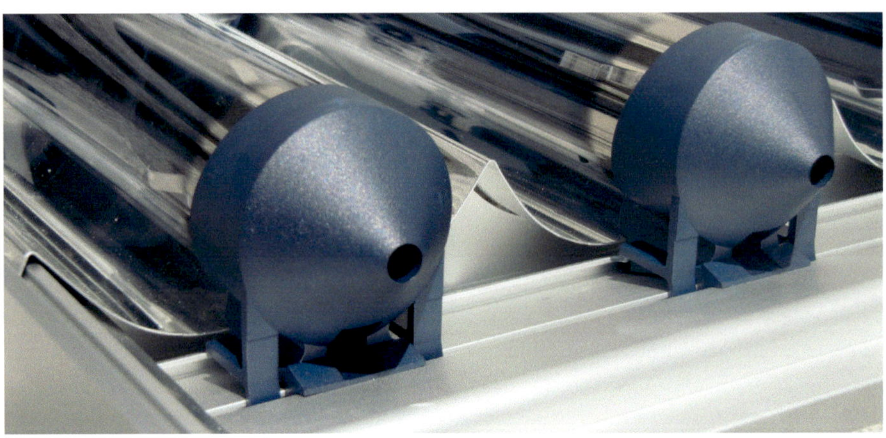

Figure 2.12 Close-up view of a CPC collector. Note the parabolic reflectors below the evacuated tubes.

Source: Solem Consulting

Table 2.4 Advantages and disadvantages of CPC collectors for solar cooling systems

Advantages	Disadvantages
• higher supply temperature than standard evacuated tube collector	• requires larger gross area on the roof • requires regular cleaning of reflectors and tubes

Parabolic trough collectors

The name says it all – parabolic trough collectors use a trough-shaped parabolic mirror to reflect the sun onto the absorber tube (see Figure 2.13). The concentration ratio of the mirror is much higher than for CPC collectors, typically between 8 and 80. The concentrated radiation results in high supply temperatures which can reach 400 °C (752 °F) or higher. Until recently, parabolic trough collectors had been used solely for power generation in large commercial solar fields. These collectors typically have an aperture width of around 6 m and provide hot thermal oil for use in a steam cycle power block. Today there are smaller parabolic troughs available for process heat and solar air-conditioning. These collectors typically have aperture widths of around 1.2–1.8 m and provide supply temperatures up to 250 °C (482 °F). See Chapter 12.7 for a supplier's list of parabolic trough collectors for solar air-conditioning. Figure 2.14 shows the view of a parabolic trough collector field.

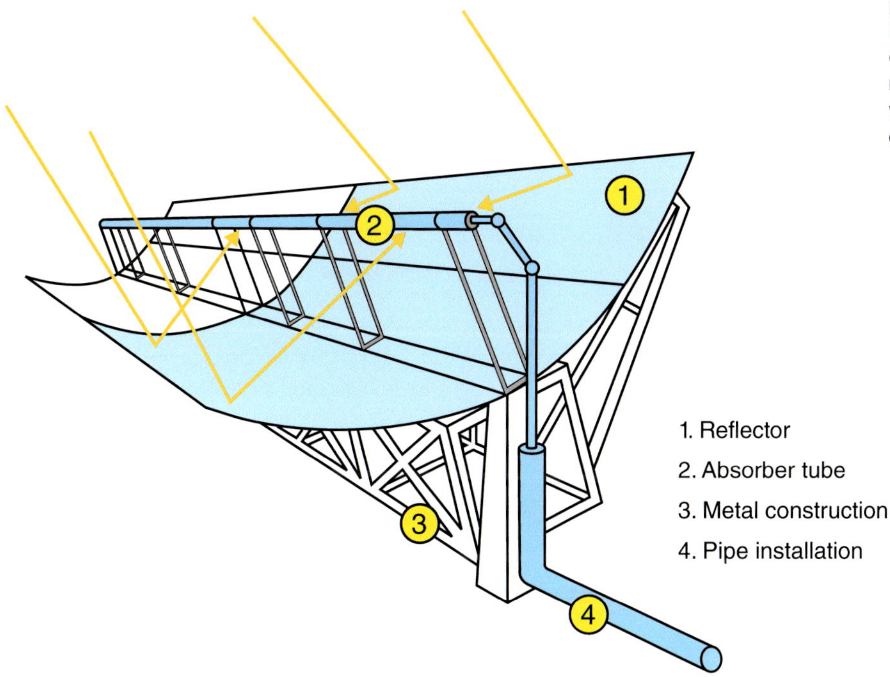

Figure 2.13 Schematic of a parabolic trough collector. Note that parabolic trough collectors require direct radiation only and will not work in areas with high diffuse radiation.

Figure 2.14 View of a parabolic trough collector field on a commercial building roof in New South Wales, Australia. The collectors provide heat for a solar cooling system.

Source: Jeremy Osborne

Table 2.5 Advantages and disadvantages of parabolic trough collectors for solar cooling systems

Advantages	Disadvantages
• higher supply temperatures than standard evacuated tube collector and CPC • higher efficiency at high supply temperature	• higher cost • not suitable for single family houses • requires regular cleaning of reflectors and tubes • requires one-axis tracking • requires direct normal irradiation (DNI)

Parabolic trough collectors require direct radiation only. Diffuse radiation cannot be reflected by the mirrors and hence will not hit the absorber tube. Parabolic troughs do not work in areas with high diffuse radiation, such as the tropics.

Linear Fresnel collectors

The linear Fresnel collector, named after French physician Augustin Jean Fresnel who invented the refractive Fresnel lens, uses multiple flat mirror strips to concentrate the sun's radiation onto the absorber tube. The mirrors are all at different angles (depending on their position relative to the absorber tube) and reflect the sunlight towards the absorber tube (see Figure 2.15). There, a secondary reflector with a parabolic shape collects the rays from the mirrors and directs them onto the absorber tube. The secondary reflector is necessary because the mirror strips do not concentrate the light themselves. They create a reflected image the same width of the mirror and typically larger than the diameter of the absorber tube. Without the secondary reflector, some of the reflected light would miss the absorber tube.

The advantage of a linear Fresnel collector is the use of multiple small mirror strips compared to the curved mirror of the parabolic troughs. The flat mirror strips are much easier to manufacture and less costly. The supporting frame can be lighter and the wind load is lower, as the mirrors are located closer to the roof/ground. However, more gross area is required compared to parabolic troughs due to the spaces between the mirrors. The absorber tube of a linear Fresnel collector is stationary. The only moving parts are the mirrors, which follow the sun using one-axis tracking.

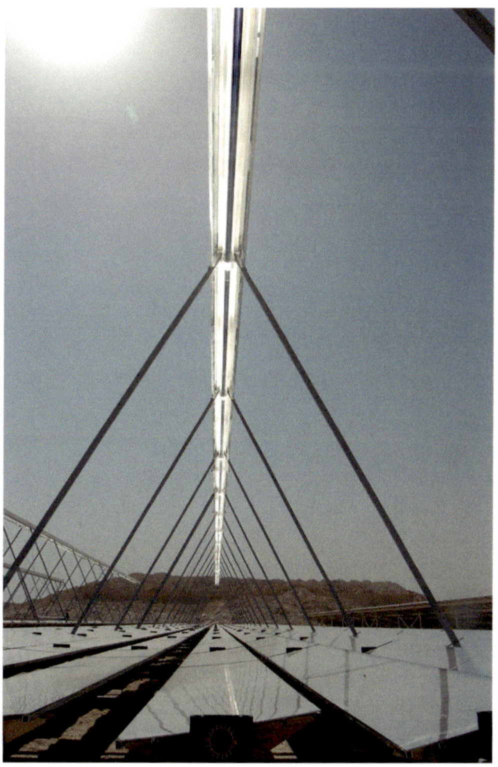

Figure 2.15 Example of a linear Fresnel collector in Qatar, UAE. The collectors supply heat to a solar cooling system for an indoor soccer stadium. Note the absorber tube in the centre illuminated by the reflected sunlight from the mirror strips below.

Source: Industrial Solar GmbH, Germany

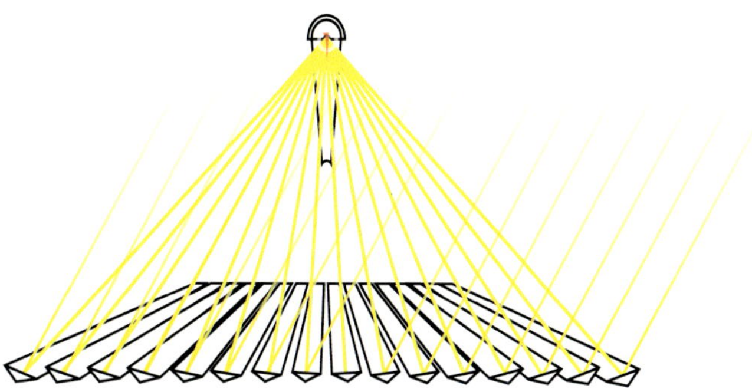

Figure 2.16 Schematic of a linear Fresnel collector.

Recently, a new small-scale product has entered the market. Called a micro-Fresnel collector, it consists of two absorber tubes integrated in a glass-covered box of trapezoidal cross-section (see Figures 2.17 and 2.18). It has been designed for use on rooftops where architectural and integration criteria play a role.

Most concentrating collectors require rather complex roof-invasive fixtures that are hard to retrofit. Also, concentrating collectors are heavier and more complex from a system point of view. Due to their shape the wind load can be much higher than for, for example, flat plate collectors, so the support structure has to be more solid. They are usually preferred for ground applications.

Note: Wind load calculation is an important issue for all solar thermal collectors. The details are beyond the scope of this book. Designers and installers should make sure that all installed systems meet local building code requirements in this regard and the issue should be discussed and clarified with suppliers of equipment. For very large systems the services of a structural engineer may be required.

Figure 2.17 Micro-Fresnel concentrator for the use on rooftops.

Source: Chromasun Inc., USA

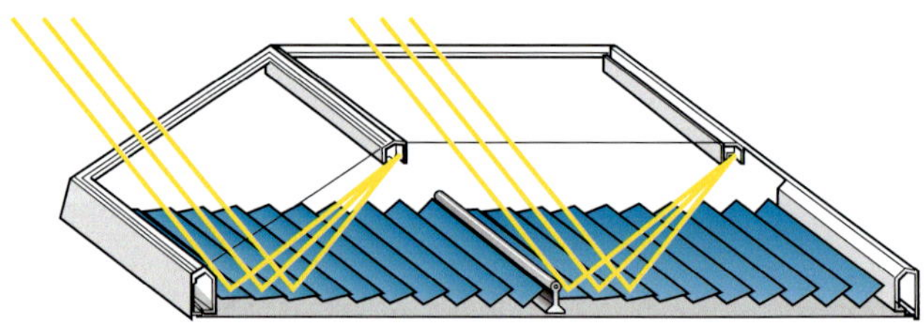

Figure 2.18 Schematic of a micro-Fresnel concentrator for use on rooftops.

Source: Chromasun Inc., USA

Table 2.6 Advantages and disadvantages of linear Fresnel collectors for solar cooling systems

Advantages	Disadvantages
• higher supply temperatures than standard evacuated tube collector and CPC • higher efficiency at high supply temperature • flat mirrors are cheaper than curved mirrors of parabolic troughs • stationary absorber tube (no moving pipe connections required) • lower susceptibility to wind forces	• lower efficiency due to secondary reflector compared to parabolic troughs • not suitable for single family houses • requires larger gross area on the roof than parabolic troughs • requires regular cleaning of mirrors and tubes • requires one-axis tracking • requires direct normal irradiation (DNI)

2.1.5 Photovoltaic-thermal collectors

Photovoltaic-thermal (PVT) collectors are hybrid collectors comprising a flat plate absorber for heat collection and a photovoltaic (PV) module for electrical power generation. A PV module converts some of the energy contained in the sunlight directly into electricity. For details about PV technology see [5]. The PV modules heat up during operation and the heat is collected in the absorber below. This cools the PV modules, thus providing a higher efficiency in power generation. PVT collectors come in two different versions: (a) unglazed and (b) glazed, i.e. with a glass cover over the PV module to reduce thermal loss. Also, PVT collectors are classified according to the heat transfer fluid through the absorber, which can be water or air. The water-based versions reach supply temperatures of maximum 70 °C (158 °F) and the air-based versions of up to 50 °C (122 °F).

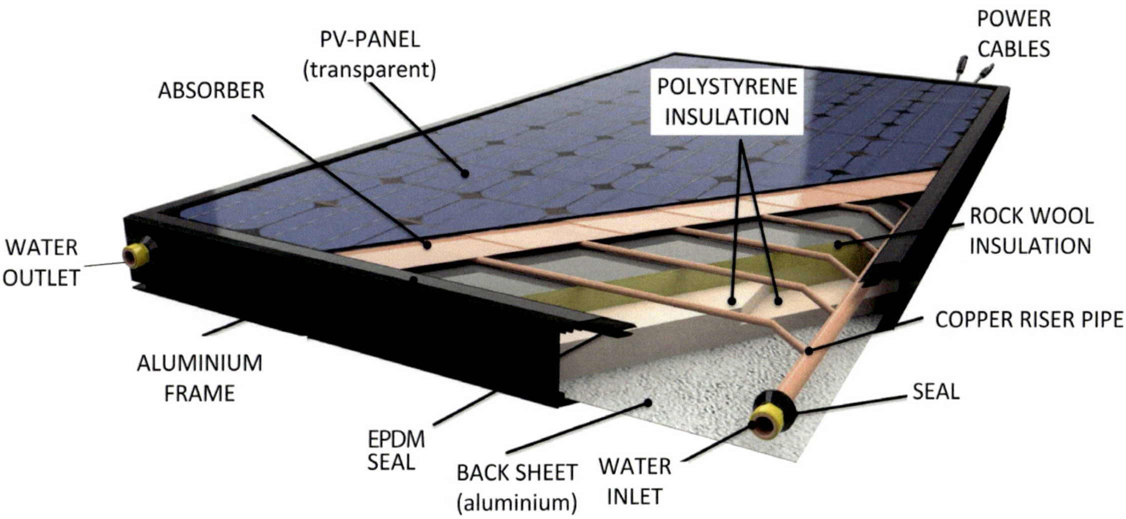

Figure 2.19 Schematic of a PVT collector.

Source: Solimpeks

Table 2.7 Advantages and disadvantages of PVT hybrid collectors for solar cooling systems

Advantages	Disadvantages
• parallel heat and power generation	• supply temperature limited due to PV cell performance • lower efficiency than standard thermal collector • transparent PV panel has lower efficiency than opaque panels • complexity of installation

2.2 Basic physics of solar thermal collectors

Solar thermal collectors are an important component of a solar cooling system. They are available in multiple technological types and options. In principle, each solar thermal collector uses an absorbing surface (the absorber) to collect the sun's radiation and transform it into heat. This heat is then transferred to a heat transfer fluid to drive the solar cooling system. The higher the amount of heat transferred to the heat transfer fluid, the better the collector performance and its efficiency. Nearly all solar collectors (with the exception of those used for pool heating) have a transparent glass cover above the absorber to prevent excessive heat loss. The shape of the absorber and the glass cover varies. It can be rectangular or cylindrical, depending on the collector type. Also, a double glass cover is sometimes used. A further classification between the collector types can be made with regard to three main technological features: concentration, tracking and vacuum. Table 2.8 shows the solar thermal collectors that can be used for solar cooling systems, classified according to the above features. Figure 2.20 shows an overview of supply temperatures and applications for the different collector types.

The efficiency of solar thermal collectors is an important parameter for the design of a solar cooling system.

Efficiency

In general, the term 'efficiency' refers to the ratio of benefit to cost of any kind of system. Applied to a solar thermal collector it is the ratio between useful heat from the collector divided by the solar energy falling onto the collector surface. It is used for instantaneous power calculations as well as annual performance simulations.

The collector efficiency η in a steady state (i.e. with no change of operating conditions) is defined as:

$$\eta = \frac{\textit{instantaneous collector power}}{\textit{incident radiation on collector area}} = \frac{\dot{Q}_{coll}}{A \cdot G} \tag{2.1}$$

where

A is the aperture or absorber area of the collector, depending on the chosen reference [m^2]

G is the incident global solar irradiance on the collector aperture [W/m^2]

$\dot{Q}_{coll}$ is the usable thermal power of the collector [W].

Table 2.8 Classification of solar thermal collectors. The supply temperature that can be achieved with the different collectors varies depends on location, radiation and collector type.

Collector name	Concentration ratio	Tracking mechanism	Absorber space	Heat transfer fluid
Air	1	None	Not evacuated	Air
Flat plate	1	None	Not evacuated[1]	Water or water–glycol
Evacuated tube	1	None	Evacuated	Water or water–glycol
Photovoltaic-thermal	1	None	Not evacuated	Water or water–glycol
Compound parabolic concentrator	1–2	None	Evacuated	Water or water–glycol
Parabolic trough	8–80	Single-axis	Evacuated[2]	Water or water–glycol, thermal oil
Linear Fresnel	30–50 (>60 using secondary reflector)	Single-axis	Evacuated	Water or water–glycol, thermal oil

Notes:

Concentration ratio is the ratio of aperture to absorber area of a solar collector. Absorber space is the volume around the absorber sheet. Tracking means actively pointing a collector towards the sun.

[1] Some flat plate collector modules with vacuum are available.

[2] Some parabolic trough collectors use a non-evacuated absorber tube.

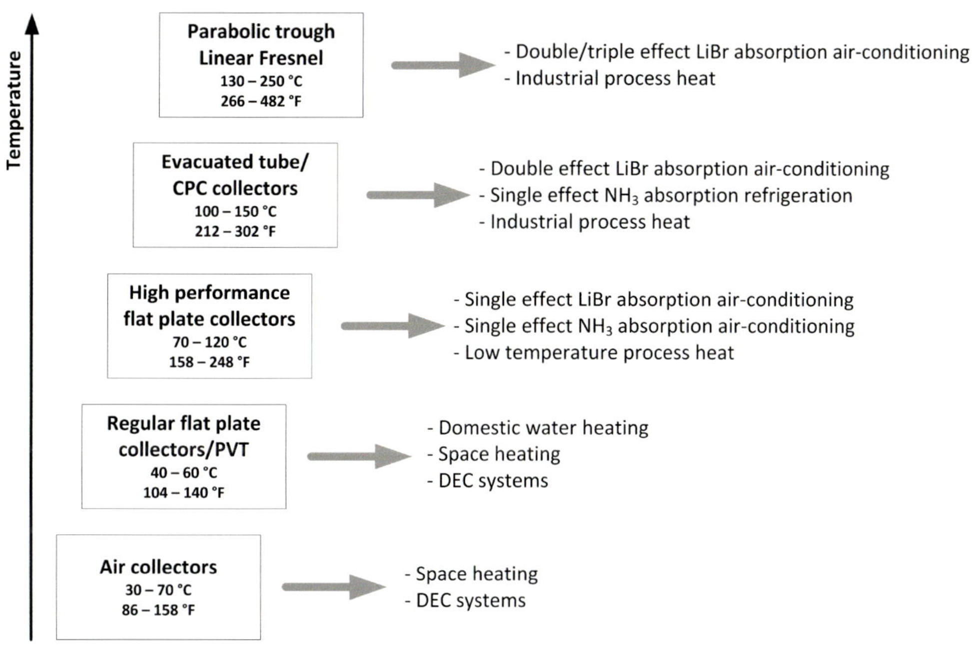

Figure 2.20 Temperature levels and applications of solar thermal collectors.

Source: Solem Consulting

Collector efficiency can also be calculated using temperature instead of collector power and aperture area:

$$\eta = \eta_0 \cdot \mathrm{k}(\Theta) - c_1 \cdot \frac{\Delta T}{G} - c_2 \cdot \frac{\Delta T^2}{G} \qquad (2.2)$$

where

η_0 is the optical collector efficiency [-]

c_1 is the linear loss coefficient [W/(m²K)]

c_2 is the quadratic loss coefficient [W/(m²K²)]

$\mathrm{k}(\Theta)$ is the incident angle modifier (IAM) [-]

ΔT is the temperature difference between mean collector temperature and ambient [K]

$$\Delta T = T_m - T_{Amb} = \frac{T_{coll,in} + T_{coll,out}}{2} - T_{Amb} \qquad (2.3)$$

where

$T_{coll,in}$ is the inlet temperature of collector heat transfer fluid [K]

$T_{coll,out}$ is the outlet temperature of collector heat transfer fluid [K].

Note that Equations (2.1) and (2.2) are for steady-state operation of the collector. It does not account for transient effects, such as the dynamic heating and cooling of the collector during operation due to its heat capacity. Also, no difference is being made between the incidence angle modifier (see Figures 2.21 and 2.22) for direct and diffuse radiation. For a more detailed description of efficiency calculations, including the above, refer to [6].

The collector coefficients η_0, c_1 and c_2 and the incident angle modifier $\mathrm{k}(\Theta)$ are usually given by manufacturers or can be found in the test reports of collector test centres. The incident angle modifier $\mathrm{k}(\Theta)$ represents the influence of the radiation incident angle on the optical performance of the solar collector. It is an important parameter that accounts for the effect of non-orthogonal light falling onto the collector and the resulting losses or gains in collector power. It is expressed as a differential change in collector efficiency compared to the efficiency value for light falling orthogonally onto the collector. For stationary or single-axis tracking collectors the IAM is per definition equal to 1 at normal incident radiation (radiation falling orthogonally onto the collector surface). The IAM is always 1 in the case of a dual-axis tracking collector (e.g. parabolic dish-type concentrators). There, the incident angle is always 90° due to the tracking. Figures 2.21 and 2.22 show exemplary IAM curves for flat plate and evacuated tube/CPC/one-axis tracking collectors.

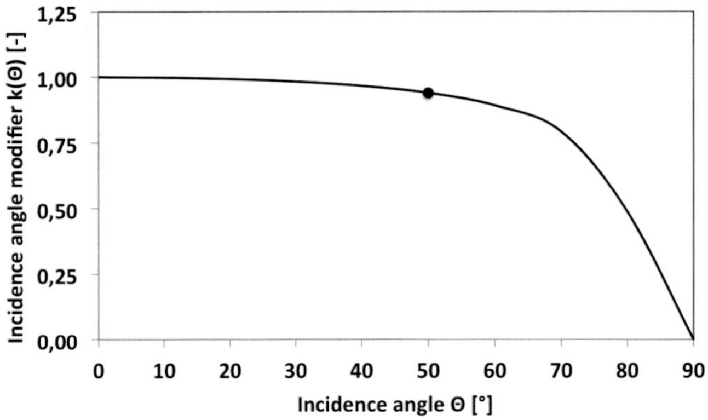

Figure 2.21 Example of incidence angle modifier k(Θ) for flat plate collectors. The IAM at 50° is typically used to describe the optical efficiency reduction at this angle.

Source: Solem Consulting

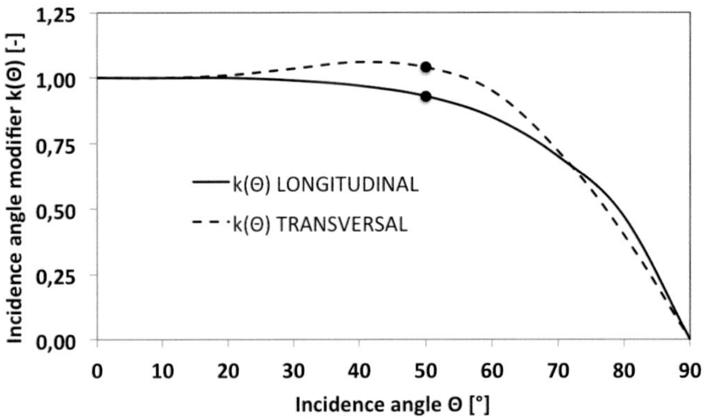

Figure 2.22 Example of incidence angle modifier k(Θ) for evacuated tube, CPC and one-axis tracking collectors. The IAMs at 50° are typically used to describe the optical efficiency reduction at this angle in both directions.

Source: Solem Consulting

For flat plate collectors the IAM k(Θ) is typically given for an incidence angle of 50° in manufacturer data or test reports (see Figure 2.21). For evacuated tube, CPC and single-axis tracking collectors two IAM values have to be provided, one to account for the longitudinal (parallel to the tube length) and one for the transversal (cross-wise to the tube length) direction (see Figure 2.22). A single IAM value for evacuated tube, CPC and single-axis tracking collectors can then be calculated using Equation (2.4):

$$k(\Theta) = k_{LONG}(\Theta) \cdot k_{TRANS}(\Theta) \tag{2.4}$$

where

k(Θ) is the total incident angle modifier (IAM) [-]

$k_{LONG}(\Theta)$ is the IAM for the longitudinal (parallel to the tube length) axis of the collector [-]

$k_{TRANS}(\Theta)$ is the IAM for the transversal (cross-wise to the tube length) axis of the collector [-]

Typical efficiency values for solar thermal collectors are given in Figure 2.23.

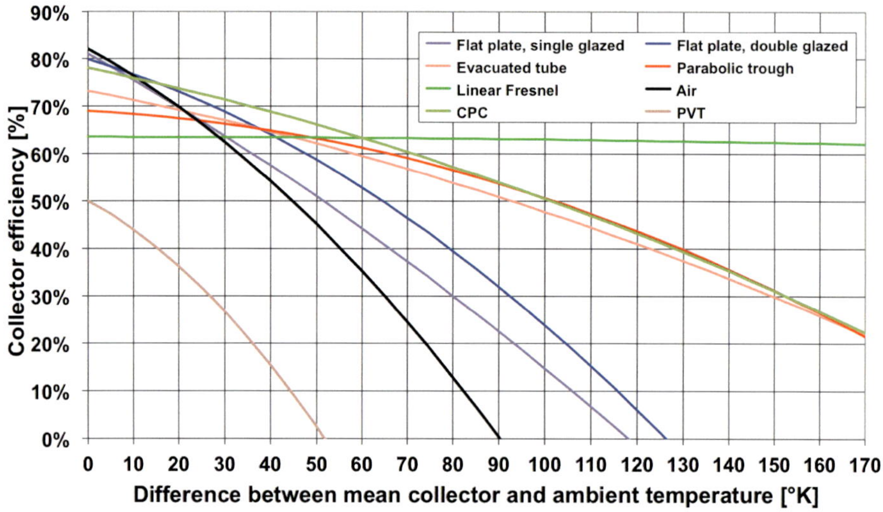

Figure 2.23 Exemplary collector efficiency graphs. All graphs are displayed for a global irradiation of 800 W/m^2 except for the parabolic trough and linear Fresnel collector, where direct normal irradiation (DNI) was used. The ambient temperature is 30 °C (86 °F) for all graphs. The linear Fresnel collector shown uses an evacuated absorber tube, the parabolic trough collector a non-evacuated absorber tube.

Source: Solem Consulting

The optical efficiency η_0 is the efficiency value for the collector operating with no temperature difference between mean collector temperature and ambient temperature (ΔT) (see Equation (2.2)). This means that the collector has the same average temperature as the immediate surrounding environment. In Figure 2.23 the optical efficiency η_0 can be found for each curve as the intersection with the *y*-axis of the graph. It can be seen in Figure 2.23 that flat plate and air collectors with rectangular absorbers typically have the highest optical efficiency; however, their efficiency drops with increasing temperature difference. Collectors using a cylindrical absorber and a vacuum around the absorber (evacuated tubes, parabolic troughs, linear Fresnel) have a lower optical efficiency but higher collector efficiencies at higher temperatures due to the increased thermal insulation provided by the vacuum. Figure 2.23 also shows the stagnation temperature of solar thermal collectors. This is the temperature where the collector efficiency is equal to zero. In this case, there is no fluid flow through the collector. It heats up due to the solar radiation until it reaches its maximum temperature. This temperature is called stagnation temperature and can reach more than 200 °C (392 °F). In Figure 2.23 the mean stagnation temperature of a collector can be calculated by adding the ambient temperature to the value of ΔT at the intersection of a collector curve with the *x*-axis. The influence of the stagnation temperature is further discussed in Chapter 7.2.1. Additional information on stagnation behaviour is also explained in [18, 19]. Some selected parameters of solar thermal collectors are given in Table 2.9.

Worked example

The efficiency η_0 of a single-glazed flat plate collector (the purple curve in Figure 2.23) is 30 per cent at a temperature difference ΔT of 80 K. For an ambient temperature of 30 °C (86 °F or 303.15 K) this results in a mean collector temperature of 303.15 K + 80 K = 383.15 K or 110 °C (230 °F).

The optical efficiency η_0 for this collector is the intersection of the purple curve with the *y*-axis, in this case approx. 80 per cent.

The mean stagnation temperature is the sum of ΔT at the intersection of the purple curve with the *x*-axis and the ambient temperature. In this case ΔT = 117 K, $T_{ambient}$ = 303.15 K, therefore $T_{stagnation}$ = 420.15 K or 147.15 °C (297 °F).

Table 2.9 Typical selected solar thermal collector parameters

Type	Optical efficiency[1] η_0	Linear loss coefficient[1] c_1	Quadratic loss coefficient[1] c_2	Aperture area[2] per module[3]	Installation area[4] per module	Specific weight[5]
	–	W/(m$_2$K)	W/(m$_2$K$_2$)	m$_2$	m$_2$	kg/m$_2$
Air collector (flat selective absorber)	0.82	4.2	0.03	2.3	2.5	36
Flat plate (selective absorber, single glazed)	0.81	4.3	0.01	1.9	2.0	17
Flat plate (selective absorber, double glazed)	0.79	2.3	0.02	2.5	2.7	30
PVT	0.50	4.1	0.07	1.4	1.5	22
Evacuated tube (direct flow)	0.73	1.5	0.005	3.0	4.1	26
CPC (Sydney tube)	0.78	1.6	0.006	1.7	2.1	24
Parabolic trough[6]	0.69	0.4	0.001	18.5	22	32
Fresnel[7]	0.64	0.0	0.0004	22	31	27

Notes:

[1] Optical efficiency and both loss coefficients are given for an irradiation of 800 W/m^2 and per aperture area.

[2] Aperture area is the net heat collecting area of a collector module.

[3] A module is the smallest quantity commercially available from the manufacturer.

[4] Installation area is the gross area required on the roof/ground for a collector module.

[5] Specific weight is per installation area and includes collector module only (no piping, etc.).

[6] The parabolic trough collector data is for a non-evacuated absorber tube.

[7] The linear Fresnel collector data is for an evacuated absorber tube.

Table 2.9 illustrates the fact that non-concentrating collectors using non-evacuated absorbers have a higher optical efficiency but lower overall collector efficiency compared to concentrators and collectors using evacuated absorbers. The linear and quadratic loss coefficients c_1 and c_2 reduce the optical efficiency at elevated temperature differences (ΔT) (see Equation (2.2)). It can be seen that c_1 and c_2 differ by an order of magnitude for, for example, single-glazed flat plate and parabolic trough collectors. Therefore, the thermal loss at elevated temperature differences is lower for the concentrating collectors using an evacuated absorber.

3

Solar cooling technologies

In this chapter we discuss the different system technologies that are available for solar cooling. The chapter includes heat-driven air-based and water-based systems, as well as electricity-driven photovoltaic (PV) systems. The physics behind these cooling technologies is also explained.

Essential solar cooling terms

ABsorption – the incorporation of a substance in one phase into another substance of a different phase (e.g. gases being absorbed by a liquid). This involves the whole volume of the liquid

ADsorption – the physical bonding of a gaseous or liquid substance onto the surface of another substance of a different phase (e.g. vapour adsorbed to a solid surface). This process involves the surface only

Chilled water – water used for process cooling or air-conditioning purposes; here, the water circuit of an AB-/ADsorption chiller with the lowest temperature

CHP – combined heat and power generation, also known as cogeneration. Refers to the process of simultaneous generation and use of heat and power, typically from a cogeneration unit. Heat is released from all thermal power generation processes (e.g. coal-fired power plants); however, it is usually rejected to the ambient and therefore lost. CHP units capture most of this heat, which increases the fuel utilization efficiency

Cogeneration – see CHP

Cooling capacity – the cooling power of a thermal chiller, typically referring to the energy removed in the chilled water circuit, in kW_r or RT

Cooling water – water used to reject the heat collected in a thermal cooling process; here, the water circuit of an AB-/ADsorption chiller with the medium temperature

District heating/cooling – a process where heat or cold is distributed from a central plant via insulated pipes to residential and commercial requirements

Evaporative cooling – the process of cooling a water body or flow via the partial evaporation of some of it. Heat required for the evaporation of water is removed from the body of water itself, hence cooling it

Hot water – water used to drive a thermal cooling or air-conditioning process; here, the water circuit of an AB-/ADsorption chiller with the highest temperature

HVAC – heating, ventilation and air-conditioning

RT – refrigeration ton; a unit for the cooling capacity of chillers. 1 RT = 3517 W_r

Refrigerant – a substance used in a heating or cooling cycle. Evaporation of the refrigerant enables the uptake of heat into the cycle. A reversible phase change is required for refrigerants

Sorbent – a substance used in a sorptive heating or cooling cycle to absorb or adsorb refrigerant vapour. Sorbents can be solids or liquids

Sorption – the collective term for AB- and ADsorption processes; see Chapter 3.3 for details

Wet bulb temperature – for given ambient conditions, the wet bulb temperature is the lowest temperature that can be reached by the evaporation of water in air

3.1 Introduction

Solar cooling technology is not a new technology, but has recently experienced large growth and interest due to worldwide energy cost increases. The hub of development and the largest markets are currently in Europe, but Asia, the USA and Australia have also been identified as significant potential markets. Solar cooling is a promising alternative to conventional vapour-compression air-conditioning, especially in areas where electricity comes at a premium cost. The peak electricity demand for air-conditioning during the summer months sometimes causes power grids to work to maximum capacity or sometimes even to fail, causing blackouts. The application of solar cooling technologies can take peak electrical loads off the grid and assist in greenhouse gas emission reduction.

In theory, the design options for a solar-driven cooling system are manifold. Both electricity and heat can be used to drive a solar cooling system. In the first case, electricity is generated from the sun using photovoltaic cells. This electricity can drive vapour-compression or thermo-electric cooling processes; these will be explained later in this chapter. In the second case, heat is generated using a variety of solar thermal collectors. The heat can drive different thermal cooling processes which are also explained in this chapter. Figure 3.1 shows an overview of common solar cooling and refrigeration technologies.

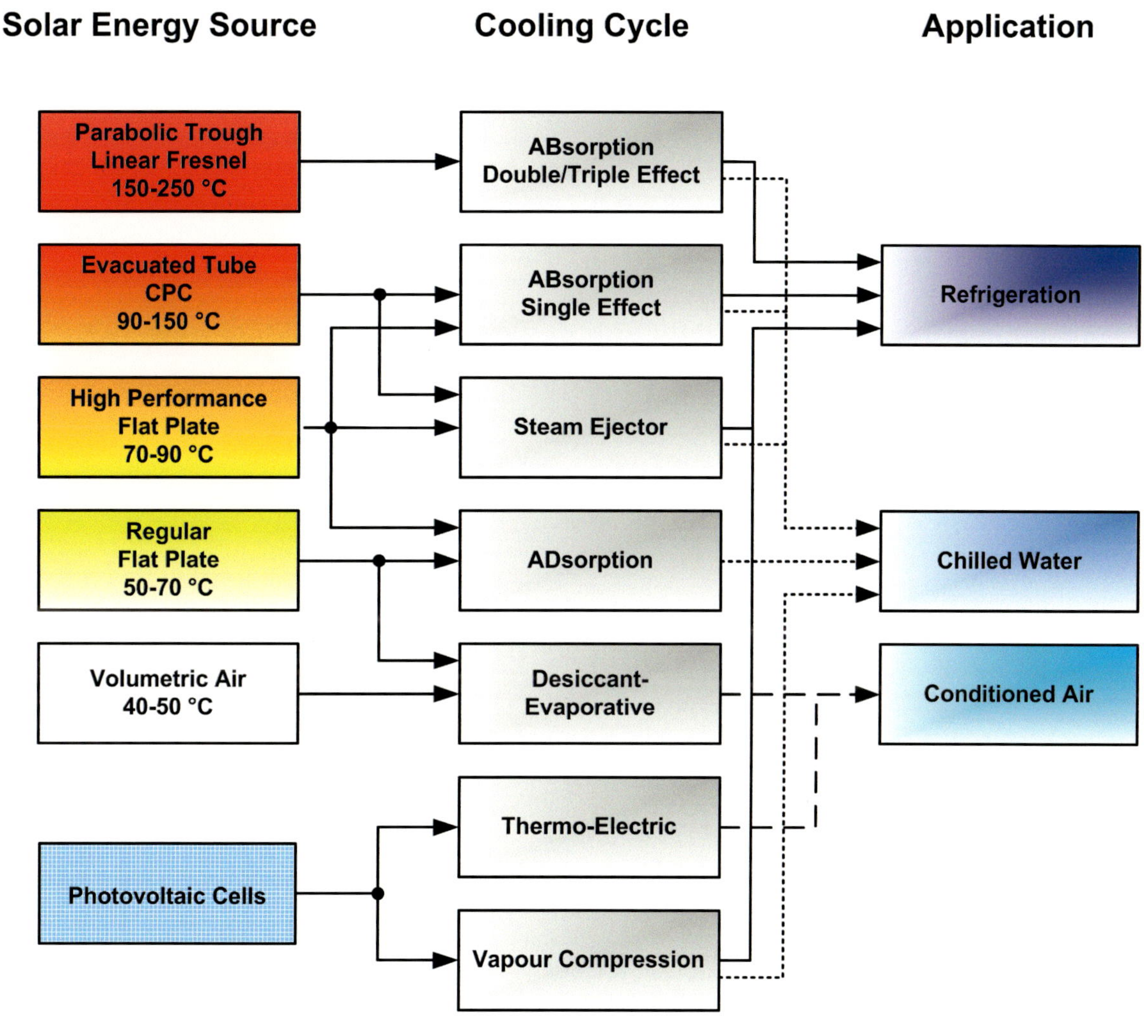

Figure 3.1 Overview of solar cooling technologies.

Source: Solem Consulting

Applications

Solar cooling can be used for the following applications:

- Refrigerating spaces to a minimum temperature of –30 °C (–22 °F)
- Chilling water to a minimum temperature of 5 °C (41 °F)
- Conditioning air (dehumidifying and chilling) to a minimum temperature of 16 °C (64 °F)

Out of all the solar-driven options shown in Figure 3.1, only four are being commercially used for building applications at present:

1. ABsorption chillers
2. ADsorption chillers
3. Desiccant-Evaporative systems (DEC)
4. PV cells with vapour-compression systems

These four systems are usually classed into two main groups, (1) open systems and (2) closed systems. These are explained in more detail in this chapter. However, some explanation of the fundamentals of cooling physics is required to understand the technologies.

Brief explanation of the three main solar thermal cooling technologies

ABsorption chillers

ABsorption chillers use a continuous cycle process based on two liquids. One liquid is the refrigerant; the other one is the carrier fluid. The refrigerant is periodically absorbed (i.e. taken up) and desorbed (i.e. released) in/from the carrier fluid, respectively. Cold is provided through the simultaneous evaporation and absorption of the refrigerant in the carrier fluid. Heat is required to desorb the refrigerant from the carrier fluid again. Cold is provided as chilled water. See Chapter 3.3.1 for details.

ADsorption chillers

ADsorption chillers use a quasi-continuous process based on a solid and a liquid. The liquid is the refrigerant; the solid is the carrier medium. The refrigerant is adsorbed in the solid while providing cold through evaporation. Once the solid is saturated with refrigerant, the process is reversed and the refrigerant is desorbed (i.e. released) from the solid. For this, heat is required. The ADsorption process is not continuous but alternates between adsorption and desorption. Cold is provided as chilled water. See Chapter 3.3.2 for details.

Desiccant-evaporative systems

Desiccant-evaporative systems (DEC) provide cold as conditioned air (i.e. air that is dehumidified and chilled). They use a solid or a liquid to remove humidity from the air. The air is then chilled using the evaporation of water. Heat is required to continuously remove the humidity taken out of the air from the liquid or solid. See Chapter 3.2 for details.

3.1.1 Fundamental physics of cooling processes

Cold is the absence of heat; therefore the effect of cooling is the removal of heat from a solid body, a liquid or gaseous flow. This is important to understand – cooling is not 'generated' or 'produced', it is simply a change of temperature caused by removing heat.

3.1.2 Conventional cooling

Conventional (i.e. non-thermal-driven, electrically-powered) cooling, air-conditioning and refrigeration is typically based on the vapour-compression cooling cycle, where mechanical energy from an electric motor drives a compressor (see Figure 3.2). The purpose of the compressor is to take the refrigerant vapour (1) to a high pressure level so that the refrigerant can reject heat at a higher temperature than it is cooling at. At high pressure, the refrigerant (2) is liquefied through a condenser using external water or air. The liquid refrigerant (3) is led back to the evaporator (4) to evaporate again and close the cycle. The heat removal from the load is via heat transfer to the liquid refrigerant, thus evaporating it. The first application of the vapour-compression cooling cycle was patented in 1834 by Jacob Perkins in London, using ether as the refrigerant.

Nowadays, cheap mass-manufacturing has led to an almost exclusive use of this cycle for heating, cooling, air-conditioning and refrigeration purposes around the globe. Common applications are house and car air-conditioners, fridges, freezers and heat pumps.

The cold temperatures of the vapour-compression cycle depend on the refrigerant being used and vary between +15 and –30 °C (59 and –22 °F). The cycle can also be operated with a focus on heat (use of the heat output by the condenser (reject heat)) and is then called a heat pump cycle. Air-conditioners are commonly sold as so-called reverse cycle units, incorporating both options of heating and cooling in one unit.

A second option to generate cooling from electricity is the thermo-electric cooler (TEC), also known as a Peltier element (named after its inventor, James Peltier). In this case, a combination of two different semiconductor materials

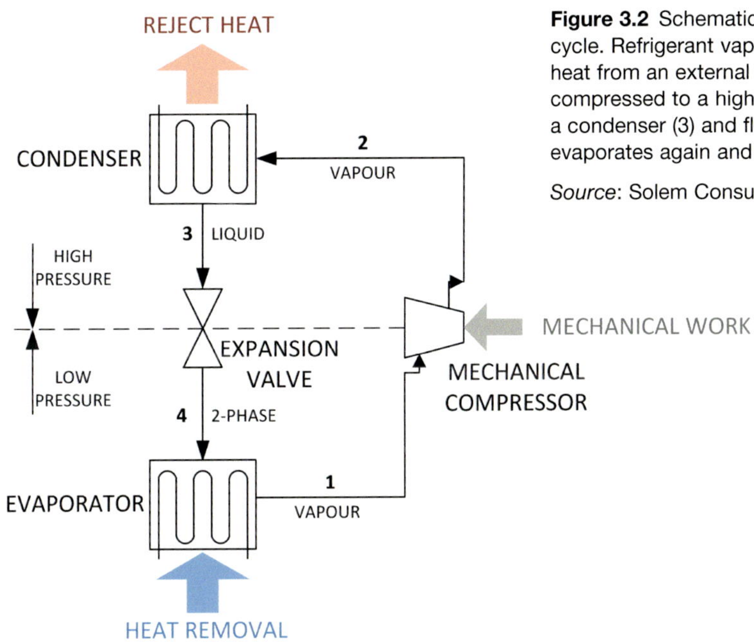

Figure 3.2 Schematic of the vapour-compression cooling cycle. Refrigerant vapour (1) is generated by removing heat from an external source (thus cooling this source) and compressed to a higher pressure (2). It is then liquefied through a condenser (3) and flows back to the evaporator (4), where it evaporates again and closes the cycle.

Source: Solem Consulting

provides cooling when a current flows through both. TEC is commercially available, but only in small sizes and cooling capacities (a few hundred watts). Therefore its use is limited to niche applications where vapour-compression systems cannot be used. These include applications with constraints on mechanical vibrations, noise or space. Examples for the use of a thermo-electric cooler are hotel mini-bar fridges, cooling of CCD chips in digital cameras and medicinal fridges.

3.1.3 Thermally driven cooling

The term 'thermally driven cooling' (sometimes called heat-driven cooling) describes the use of various heat sources to drive a thermal cooling process. Heat sources include, but are not limited to, solar energy, district heat, reject heat from combined heat and power (CHP) units or waste heat from industrial processes. Thermal cooling processes employ ozone- and climate-friendly refrigerants and have a rather low electricity demand compared to vapour-compression units. Hence their CO_2 emission balance is considerably better. A thermal cooling process driven by solar thermal energy is called solar air-conditioning, solar cooling or solar refrigeration, depending on the application. The main advantage is the coincidence of solar irradiation and cooling demand in summer months, leading to low electricity consumption for the cooling provided.

The two main physical principles of thermally driven cooling processes are

- evaporation of a refrigerant by means of heat uptake from a separate liquid flow, thus cooling the liquid flow;
- evaporation of water by means of heat uptake from an air flow, thus cooling the air flow (also known as evaporative cooling).

Thermally driven cooling technologies have been used for over 100 years, but there are still only a handful of technologies commercially available. These can be divided into systems to provide chilled water and systems to provide conditioned (chilled and dehumidified) air (see Figure 3.3).

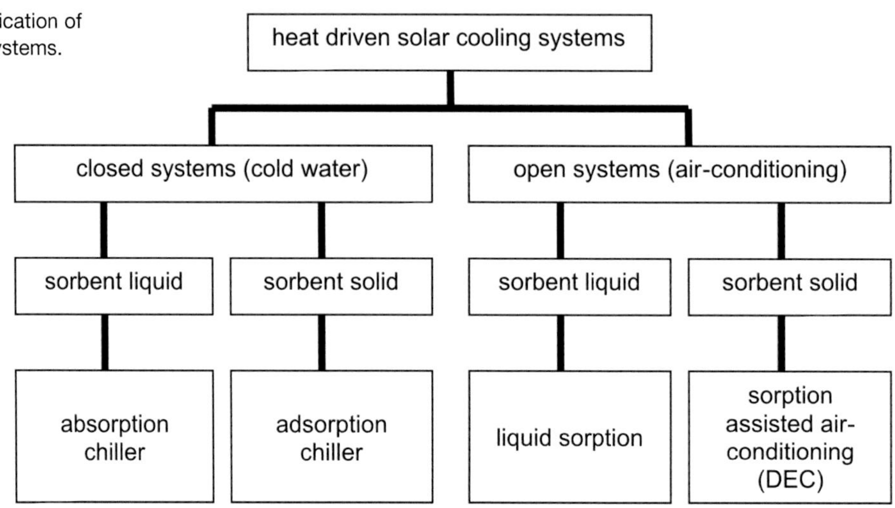

Figure 3.3 General classification of thermally driven cooling systems.

Source: Solem Consulting

Table 3.1 shows the working pairs, i.e. the combination of refrigerant and sorbent, and the cooling medium, as well as the different temperature ranges for chilled, heated and cooling water, the cooling capacity per unit and the coefficient of performance (COP) of commercially available solar thermal cooling technologies.

Table 3.1 Solar thermal driven cooling and air-conditioning technologies

	ABsorption			ADsorption	DEC
	Single-effect	Double-effect	Single-effect		
Refrigerant	Water	Water	Ammonia	Water	–
Sorbent	Lithium bromide	Lithium bromide	Water	Silica gel	Silica gel or lithium chloride
Cooling medium	Water	Water	Water + glycol	Water	Air
Chilled water temperature	6–20 °C	6–20 °C	–30–+20 °C	6–20 °C	16–20 °C
Hot water temperature	75–100 °C	130–160 °C	75–150 °C	55–100 °C	55–100 °C
Cooling water temperature	25–35 °C	25–35 °C	25–50 °C	25–45 °C	Not required
Cooling capacity range (per unit)	10–20,500 kW	170–23,300 kW	19–1,000 kW	10–490 kW	6–300 kW
Coefficient of performance (COP)	0.6–0.7	1.1–1.4	0.5–0.7	0.5–0.65	0.5–1.0

3.2 Cooling with heat: air-based systems

A solar cooling system that provides conditioned air is called an open system. The principle of open systems is to employ ambient air (or a combination of recirculated building air mixed with ambient air) to air-condition a building. The air is drawn from ambient, then conditioned and delivered into the building. Building heat is removed through airflows which are controlled to a certain temperature and humidity. Because the air usually flows through the building only once, these systems are called open systems; new air is continuously drawn into the building. Open systems therefore also provide building ventilation at the same time as air-conditioning.

Within the solar-driven open systems there are two commercially available technologies, (1) solid-based and (2) liquid-based.

3.2.1 Solid DEC technology

The solid-based open systems are called desiccant-evaporative cooling (DEC) systems. Here, the ambient air is first dehumidified using a solid hygroscopic adsorption material such as silica gel, zeolite or similar. Then the evaporative cooling effect of water is used to reduce the temperature of the ambient air to the required level. Figure 3.4 shows a DEC system schematic.

A DEC cycle combines three different changes in process air conditions performed in series by the following system components:

1. dehumidification of the process air in a rotating dehumidification wheel (also known as a dehumidifier, desiccant wheel or sorption wheel);
2. pre-cooling of the process air while passing a rotating heat exchange wheel (also known as a heat recovery wheel or HX wheel);
3. evaporative cooling of the process air in evaporative coolers (also known as humidifiers or direct evaporative coolers).

In Figure 3.4 the process steps applied to the supply air (bottom airflow) side are depicted from left to right:

1. Filtering of ambient air (dust, pollen etc.).
2. Dehumidification of ambient air in the dehumidification wheel. This step removes moisture from the air to enhance the effect of the subsequent evaporative cooling step. However, as a result of moisture adsorption onto the desiccant wheel, heat is released and the air is warmed.
3. Pre-cooling of the process air while passing through a heat exchange wheel. Heat added during the previous dehumidification step is partially removed.

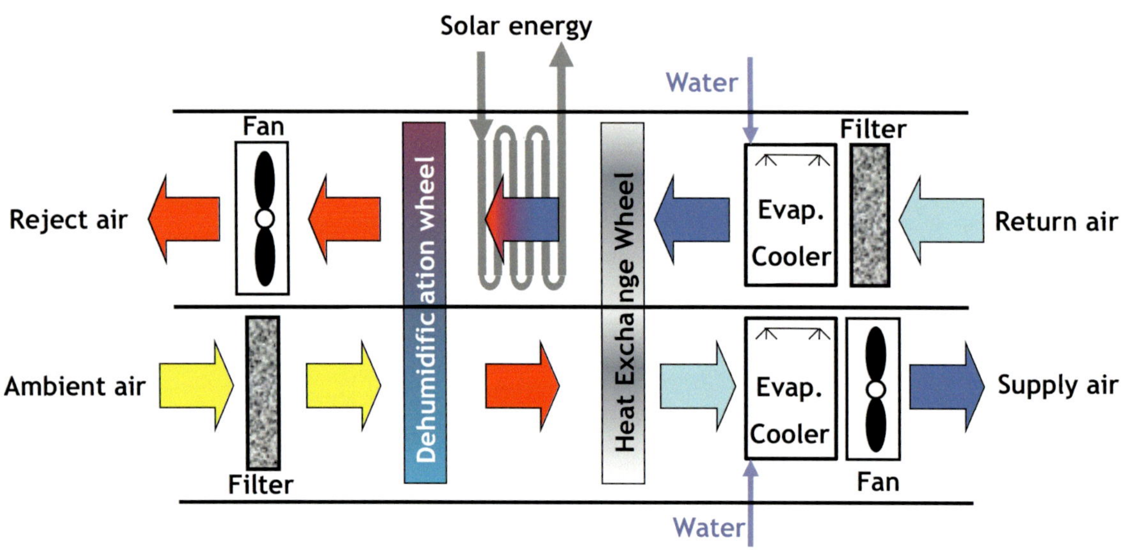

Figure 3.4 Process schematic of a solid DEC system (in cooling mode). Ambient air is drawn in from the environment. Supply air is cool, dehumidified air entering the building. Return air is drawn from the building. It is hotter and more humid than the supply air. Return air is, however, still colder than ambient and can thus be utilised in the DEC process. Reject air is hot and humid air that is exhausted to ambient.

Source: Solem Consulting

4. Evaporative cooling of the process air in evaporative coolers (direct evaporation). Evaporative cooling reduces the temperature of the supply air to the desired condition for the building.

The desiccant wheel adsorbs moisture during the above process, which has to be removed continuously to keep the process operating. The rejection of moisture is called regeneration and takes place in the top flow channel, using return air from the building. In Figure 3.4 the process steps on the return air side (top airflow) are depicted from right to left:

1. Evaporative cooling of the building return air. The return air is hotter than the supply air since it has taken up some of the building heat load. It is also usually more humid than the supply air. The evaporation of water into the return air lowers the temperature. This step maximises the cooling effect on the supply air that can be achieved in the heat exchange wheel. In combination with the heat exchange wheel, the evaporative cooler acts as an indirect evaporative cooler of the supply air.
2. Pre-heating of the return air while passing through a heat exchange wheel. Cooling of the supply air provides heat to the regeneration air, which reduces the need for additional external heat to achieve regeneration of the desiccant wheel.
3. Heating of the return air using a solar heat source. Solar heat is supplied to bring the regeneration air up to the required temperature for regeneration of the desiccant wheel.
4. Regeneration of the desiccant wheel. Hot return air drives the adsorbed moisture from the bottom flow channel off the desiccant so that the desiccant can be used again for dehumidification. The rotation of the dehumidification wheel causes continuous dehumidification and regeneration in the bottom and top channel, respectively.

DEC systems allow both controlled air cooling and humidifying at the same time. DEC technology is primarily used in buildings with high ventilation rates and fresh air demand. It is a robust technology that operates at atmospheric pressure throughout the system. A solid DEC system consists of commercially available fans, motors and water circulation pumps. The main specialist

Figure 3.5 Different sorption wheels for solid DEC systems.

Source: U. Schürger

components are the desiccant wheel and heat recovery wheel, both of which are mature, commercially available technologies taken from the dehumidification industry. The minimum supply air temperature that can be achieved is about 16 °C (60 °F). The solid desiccant cooling cycle can operate satisfactorily at heat source temperatures of 60–70 °C (140–160 °F), which is lower than closed system absorption and adsorption chillers. Examples of dehumidification wheels are given in Figure 3.5.

3.2.2 Liquid DEC technology

Liquid desiccant systems are similar to solid desiccant systems; however, the sorption component is a hygroscopic solution instead of a solid. Common sorption materials are lithium chloride (LiCl), lithium bromide (LiBr), triethylene glycol ($C_6H_{14}O_4$) and aqueous calcium chloride ($CaCl_2$). There are two methods of liquid desiccant dehumidification, applying the liquid onto a carrier matrix (e.g. a rotating wheel), or by spraying the liquid directly into the air stream. The hygroscopic solution attracts water from the air, gets diluted and is regenerated afterwards using an external heat input. Figure 3.6 shows a schematic of a liquid cooling process.

The process steps in Figure 3.6 are depicted from left to right:

1. Filtering ambient air (dust, pollen, etc.).
2. Dehumidification of ambient air in the absorber. This step removes moisture from the air through the attraction of water in a concentrated hygroscopic solution. The moist ambient air is brought into contact with the solution and moisture is removed. The removed moisture is added to the solution (hence diluting it), which is then collected in a catch tank. Similar to the DEC cycle, heat is released during the water removal and the air is warmed.
3. The supply air is cooled while passing through a heat exchanger. Heat

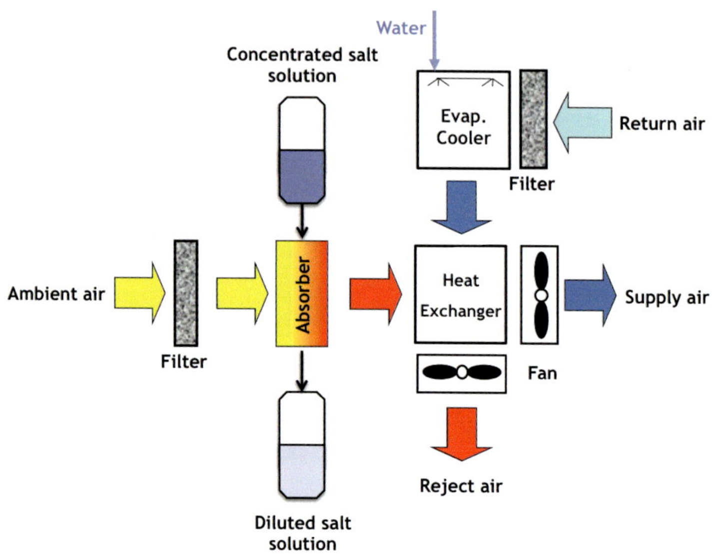

Figure 3.6 Process schematic of a liquid DEC system (in cooling mode). Ambient air is drawn in from the environment. Supply air is cool, dehumidified air entering the building. Return air is drawn from the building. It is hotter and more humid than the supply air. Return air is, however, still colder than ambient and can thus be utilised in the DEC process. Reject air is humid air that is exhausted to ambient.

Source: Solem Consulting

released into the supply during the previous dehumidification step is removed using evaporatively cooled return air from the building in a separate airflow. The supply air does not come into contact with the return air.

4. Regeneration of the diluted salt solution using solar heat – in analogy to the DEC cycle. The solution is heated and the water contained is boiled off (not shown). This step is necessary in order to restore the function of the solution for moisture removal. If too much water is contained in the solution then no more water can be attracted from the ambient air. The regeneration process takes place in a separate component that is not shown in Figure 3.6. The heat input for boiling off the water from the solution is taken from solar collectors. After boiling off the water, the solution is concentrated again.

An advantage of a liquid desiccant system compared to a solid desiccant system is that the solution can be pumped and also be stored for later regeneration at a higher storage density (the storage density of lithium chloride solution is approximately five times higher than for hot water). Some solutions have lower regeneration temperatures than solid materials (e.g. $CaCl_2$ at 41 °C); this allows the use of cost-effective solar collectors with lower efficiencies. Liquid desiccant systems are also capable of removing airborne pollutants. However, liquid desiccant systems are more complex than solid desiccant systems. The solutions used in liquid desiccant systems are usually corrosive, resulting in higher maintenance requirements. Care has to be taken that none of the corrosive solution enters the supply airflow.

A liquid DEC system consists of commercially available fans, motors and solution circulation pumps. The main specialist components are the absorber and the heat exchanger. Both components are considered technologically

Figure 3.7a Example of a liquid desiccant cooling system using lithium chloride solution. Supply air with a temperature of 24 °C/75 °F can be provided with heat source temperatures for regeneration between 70 and 80 °C (158–176 °F). The absorber plates (in blue) are clearly visible.

Source: M. Peltzer/L-DCS Technology GmbH

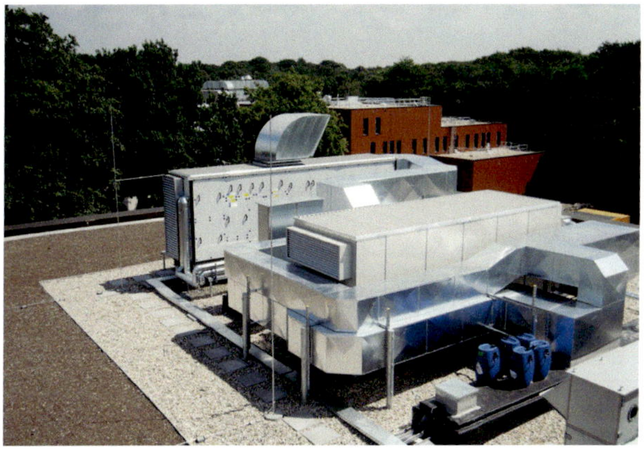

Figure 3.7b Example of a liquid desiccant system installed in a low-energy office building in Duisburg, Germany. Low-temperature district heat is used to drive the air-conditioning system.

Source: Menerga

mature; however, unlike those for solid DEC technology, they are not commercially available from the air-conditioning industry.

The minimum supply air temperature that can be achieved in a liquid DEC system is about 16 °C. The liquid desiccant cooling cycle can operate satisfactorily with heat source temperatures of 45–70 °C (113–160 °F), which is lower than of closed system absorption and adsorption chillers. Examples of commercial liquid cooling systems are given in Figure 3.7.

3.3 Cooling with heat: water-based systems

A system that uses water to remove the building heat is called a closed system. The heat removal from the building is via heat transfer into the water, which is then cooled in a water chiller. Air can be used to transfer the building heat into the water; however, air is not directly used for heat removal from the building. The building heat is rejected from the water chiller to the external environment. In a closed system, no air is drawn into the building for cooling purposes, therefore ventilation of the building has to be provided separately. As opposed to open systems, in closed systems the water is circulated within the building and not consumed.

The water chiller is the main component in a closed system. It is usually a heat-driven ABsorption or ADsorption chiller and has the same function as a conventional vapour-compression chiller – provide chilled water. The chiller usually provides cold water at temperatures between 6 °C and 20 °C (42–68 °F). Water chillers can be used for central air-conditioners as well as cooling systems with decentralised air treatment, such as fan coils and cooled ceilings. The two main technologies employed in solar-driven closed systems are AB- and ADsorption chillers.

3.3.1 ABsorption chillers

The working principle of an ABsorption chiller is based on different boiling temperatures of refrigerant and a liquid sorption medium, the absorbent. External heat input causes a refrigerant to evaporate. The refrigerant vapour is absorbed by the absorbent, thus diluting it. Pumped to a higher pressure level, external heat input evaporates the refrigerant again while the concentrated absorbent stays liquid. The absorbent then flows back to the absorber, whereas the refrigerant vapour condenses, passes an expansion device and evaporates again, starting the cycle all over again. Sources of detailed descriptions on the functioning of ABsorption chillers can be found in Chapter 12.

There are two main working pair combinations in commercial use: (1) water plus lithium bromide; and (2) ammonia plus water. Table 3.2 shows an overview of typical ABsorption chiller parameters.

Single-effect ABsorption chillers have two internal pressure levels, while double-/triple-effect chillers have three/four internal pressure levels, respectively. This allows higher heating temperatures and thus improved efficiency compared to single-effect chillers. The water/lithium bromide chillers are limited to cooling temperatures above the freezing point of water. The

Table 3.2 Overview of ABsorption chiller parameters

	ABsorption			
	Single-effect	**Double-effect**	**Triple-effect**	**Single-effect**
Refrigerant	Water	Water	Water	Ammonia
Sorbent	Lithium bromide	Lithium bromide	Lithium bromide	Water
Cooling medium	Water	Water	Water	Water–glycol
Chilled water temperature	6–20 °C	6–20 °C	6–20 °C	-30–+20 °C
Hot water temperature	75–100 °C	130–160 °C	200–250 °C	75–150 °C
Cooling water temperature	25–35 °C	25–35 °C	25–35 °C	25–50 °C
Cooling capacity range (per unit)	10–20,500 kW	170–23,300 kW	100–1,200 kW	19–1,000 kW
Coefficient of performance (COP)	0.6–0.7	1.1–1.4	1.7–1.8	0.5–0.7

Figure 3.8 An overview on commercially available ABsorption chillers is given in Chapter 9.3.

Figure 3.8a Example of a 15 kW$_r$ single-effect absorption chiller.

Source: EAW GmbH

Figure 3.8b Example of a 233 kW$_r$ double-effect absorption chiller.

Source: Solem Consulting

Figure 3.8c Example of a 563 kW$_r$ triple-effect absorption chiller.

Source: Kawasaki KTE

ammonia/water ABsorption chillers can generate cooling temperatures down to –20 °C, which is suitable for industrial cold and refrigeration processes. ABsorption chillers are commercially available with cooling capacities ranging from a few kilowatts up to megawatts. Figure 3.8 shows selected examples of ABsorption chillers.

ABsorption chillers generally require a wet cooling tower to provide heat rejection at the lowest possible temperature. This improves efficiency, maximises capacity utilisation and reduces the possibility of ABsorbent crystallisation in the case of the water/lithium bromide solution pair. The ammonia/water pair does not have crystallisation problems and can operate at higher heat rejection temperatures and is therefore more amenable to air cooling.

Maintenance of water/lithium bromide solution chillers requires an occasional purge of air from the system, which results from operating under vacuum. Furthermore, the heat exchangers to the external circuits have to be cleaned regularly to clear internal fouling. The concentration of additives in the solution needs regular checks and refills when required. Ammonia/water chillers operate at positive pressure and do not require a regular vacuum purge. Their heat exchangers also have to be cleaned regularly from internal fouling. Furthermore, ammonia is classified as toxic and such chillers are therefore subject to the relevant safety standards in the country of application.

3.3.2 ADsorption chillers

ADsorption chillers have a similar working principle to ABsorption chillers, although the sorption component is solid. Also, the cycle is not a continuous one but consists of two phases. In the evaporation phase, refrigerant steam is adsorbed by the adsorbent, thus saturating it. The adsorbent has to be cooled during this phase. In the condensation phase, the adsorbent is heated and the refrigerant evaporates again. The refrigerant steam cools and condenses while releasing heat. The liquid refrigerant is led back into the same unit where it evaporated before, and the cycle begins again. In an ADsorption chiller two such cycles run simultaneously with opposite operation modes. This ensures a continuous chilled water supply with fluctuating chilled water temperature. Sources of detailed descriptions on the functioning of ADsorption chillers can be found in Chapter 12. Figure 3.9 shows selected examples of ADsorption chillers, and Table 3.3 shows an overview of typical ADsorption chiller parameters.

Wet cooling towers (see Chapter 3.3.3) are generally used with ADsorption chillers to maximise cooling capacity, even though crystallisation is not a problem. Commercially available ADsorption chillers do not contain toxic or harmful substances. They are bulkier and heavier than ABsorption chillers. While there are a number of commercial units available and the technology is relatively mature, ADsorption chillers are far less common than ABsorption chillers.

The use of water as a refrigerant requires operation under vacuum and associated air purging as part of the maintenance routine. The heat exchangers of the external circuits require regular cleaning from internal fouling.

A comprehensive list of all commercially available AB- and ADsorption chillers at the time of printing can be found in Chapter 9.3.

Figure 3.9 An overview of commercially available ADsorption chillers is given in Chapter 9.3.

Figure 3.9a Example of Silicagel ADsorption chiller with 10 kW_r cooling capacity.

Source: Sortech AG

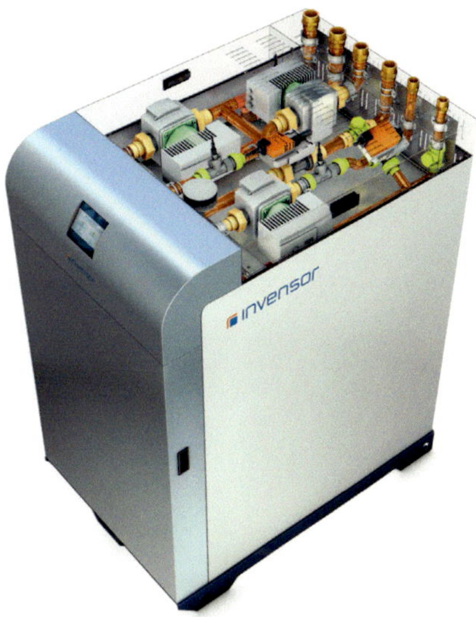

Figure 3.9b Example of Zeolithe ADsorption chiller with 18 kW_r cooling capacity.

Source: InvenSor GmbH

Table 3.3 Overview of ADsorption chiller parameters

	ADsorption	
	Single-effect	**Single-effect**
Refrigerant	Water	Water
Sorbent	Silica gel	Zeolithe
Cooling medium	Water	Water
Chilled water temperature	6–20 °C	6–20 °C
Hot water temperature	55–100 °C	45–95 °C
Cooling water temperature	22–45 °C	20–45 °C
Cooling capacity range (per unit)	10–490 kW	10–370 kW
Coefficient of performance (COP)	0.5–0.6	0.5–0.6

3.3.3 Heat-rejection technologies

Heat rejection is necessary for both AB- and ADsorption chillers. The amount of heat needing to be rejected by the system consists of both the heat to be removed from the building or space being cooled (e.g. the cold production via chilled water) plus the driving heat (e.g. the hot water heat from the solar system). The rejected amount of heat is larger for thermal cooling systems than for vapour-compression systems. Typical technologies used for the heat rejection are given in Table 3.4. Figure 3.10 shows selected examples.

Table 3.4 Overview of heat-rejection technologies for solar thermal cooling systems

Heat-rejection technology	Description
Open-loop evaporative cooling (using mains water)	Water is sprayed over the top of a fill package inside the tower and trickles to the bottom. Air is drawn through the fill package from the bottom to the top via a fan. Some water evaporates, thus cooling the remaining water. The cooled water is collected at the bottom of the tower (in the sump) and returned to the solar cooling system. Evaporated water is periodically replaced with mains water.
Closed-loop dry cooling (using air)	Water flows through a heat exchanger. Heat is rejected into air being drawn over the heat exchanger via a fan. The water does not come into contact with ambient air.
Hybrid cooling (using both mains water and air)	Combination of open-loop (evaporative) and closed-loop (dry) cooling. In cool weather conditions it operates dry, i.e. without contact between water and air. In hotter conditions it operates in evaporative cooling mode.
Adiabatic cooling (using both mains water and air)	Modification of closed-loop dry cooling using a water spray system. Water is sprayed into the air drawn over the heat exchanger and evaporates, thus providing lower cooling temperatures.
Direct cooling (using ground water or swimming pools)	Use of water that is available at sufficiently low temperatures, e.g. ground water through boreholes or pool water.

Figure 3.10 Selected examples of heat rejection devices: wet cooling tower (left); hybrid cooler (centre); dry cooler (right).

Sources: Baltimore Aircoil International (left), Güntner (centre and right)

Sources of further information on heat-rejection systems can be found in Chapter 12.3.

3.4 Cooling with heat: system comparison and overview

An overview of open and closed solar cooling technologies based on existing products and systems has been summarised in Table 3.5. Please refer to the resources section (Chapter 12.6) for a list of manufacturers. A list of commercially available small- and medium-scale chillers at the time of printing can be found in Chapter 9.3.

Table 3.5 Comparative overview of available solar cooling technologies

Technology	ABsorption chiller (single stage)	ADsorption chiller	Solid desiccant evaporative	Liquid desiccant evaporative
System type	Closed	Closed	Open	Open
Abbreviation	AB	AD	SDEC	LDEC
Heat source temperature	80 °C	60 °C	65 °C	70 °C
Cooling delivery method	Chilled water via fan coils	Chilled water via fan coils	Direct air supply	Direct air supply
Heat-rejection requirement	Wet: full performance; dry: 50 per cent reduced performance	Wet: full performance; dry: 50 per cent reduced performance	None	Depends on manufacturer/system[1]
Water consumption	Wet: 2.5 l/hr per kW cooling capacity; dry: none	Wet: 2.5 l/hr per kW cooling capacity; dry: none	3 l/hr per kW cooling capacity	3 l/hr per kW cooling capacity
Parasitic electrical power consumption (cooling only, no chilled water/air distribution)	12–20 W per kW cooling capacity	1–2 W per kW cooling capacity	25–30 W per kW cooling capacity	90–110 W per kW cooling capacity
COP (best operation)	0.8	0.65	0.7	1.0
Cold supply temperature	>4 °C (LiBr), <0 °C (NH_3)	>4 °C (LiBr)	Wet bulb temperature limit (15 °C)	Wet bulb temperature limit (15 °C)
Working fluid hazards	LiBr corrosive and irritant, NH_3 toxic	None	None	CaCl and LiCl corrosive and irritant to skin and eyes
Maintenance requirements	Regular air purge with vacuum pump, corrosion inhibitor	Regular air purge with vacuum pump, valve check	Wheel drive belts, air filters, fans	Pumps, air filters, fans, corrosion check
Weight	30 kg/kW cooling capacity	40 kg/kW cooling capacity	25 kg/kW cooling capacity	110 kg/kW cooling capacity
Part load performance	70–100 per cent	70–100 per cent	0–100 per cent (free evaporative cooling can be used at low load)	0–100 per cent (free evaporative cooling can be used at low load)
Technical maturity	Very good	Good	Moderate (good at a large scale)	Moderate
Ideal application	Radiant cooling at elevated chilled water temperature (15 °C)	Radiant cooling at elevated chilled water temperature (15 °C)	Humid environments, ventilated spaces	Humid environments, ventilated spaces
Other comments	Lower chilled water temperature needed for dehumidification (6 °C) reduces performance	Lower chilled water temperature needed for dehumidification (6 °C) reduces performance fluctuating chilled water temperature	Performance limited by climate, positive pressure in house will eliminate heat infiltration	Performance limited by climate, positive pressure in house will eliminate heat infiltration

Note:

[1] Some manufacturers require external cooling towers for their liquid sorption systems, others don't. If an external cooling tower is used then the same conditions as for closed systems apply to the heat rejection requirements.

3.5 Cooling with solar electricity: photovoltaic systems

Essential terms for cooling with photovoltaics

Inverter – a device that converts direct current (DC) output from a PV module into alternating current (AC)

Module – a PV module (sometimes also called panel) is an assembly of PV cells. It provides DC when exposed to sunlight

Feed-in tariff (FiT) – a billing mechanism where renewable energy electricity generators are being paid a fixed price for the electricity exported to the grid. Typically, contracts are long term, ensuring manageable risks for investors in renewable energy technologies.

3.5.1 Introduction

Recent developments in the PV sector, as well as growth in the air-conditioning sector, have led facility managers to consider alternatives to conventional and thermal air-conditioning. A promising concept is PV-based air-conditioning – the coupling of PV modules with an electrically driven air-conditioner. The idea behind it is relatively simple and all components are commercially available. Until recently, the high cost of PV modules has been prohibitive for the use of this technology to power air-conditioning. Recent price drops of PV modules have changed this. In 2010 a module price (excluding installation) of \$3.50/W_p reflected an approximate global average; in 2012 \$1.36/W_p (excluding installation) was found to be the current market cost [7]. This is a reduction in module cost of 63 per cent over the course of two years. Therefore, PV cooling now poses an alternative to thermal cooling.

There are two main options when using PV-driven cooling systems:

1. *Grid-connected systems* with PV power being exported to the grid and a standard grid-powered air-conditioner, as in a standard grid-connected PV system.
2. *Grid-independent systems* with battery storage and a DC/AC inverter.

3.5.2 Grid-connected PV cooling systems

Standard grid-connected PV systems – with PV power being exported to the grid and a standard grid-powered air-conditioner – are metered differently (net-metering or FiT). It is important to understand the difference.

Net-metering situations

A standard grid-connected PV system in a building generates power that is either (1) being used to substitute power purchased from the grid or (2) being exported to the grid. Such a system is referred to as net-metering, since the meter in the building accounts only for the difference between building consumption and PV power generation. This means that the meter will in fact spin backwards if PV power generation exceeds building power consumption. If an air-conditioning unit is present in the building then this is considered a grid-connected PV cooling system (Figure 3.11). However, PV cooling is only possible when the sun is shining, At night-time the air-conditioning unit operates on grid power.

The subsystems of a grid-connected PV-driven cooling system are:

1. power generation – PV modules plus inverter and electrical connections;
2. cold generation – a vapour-compression air-conditioner or reversible heat pump.

The technologies used in both of these subsystems are mature, and have been in use for many years. A grid-connected PV cooling system as described in Figure 3.11 is technically possible and commercially available.

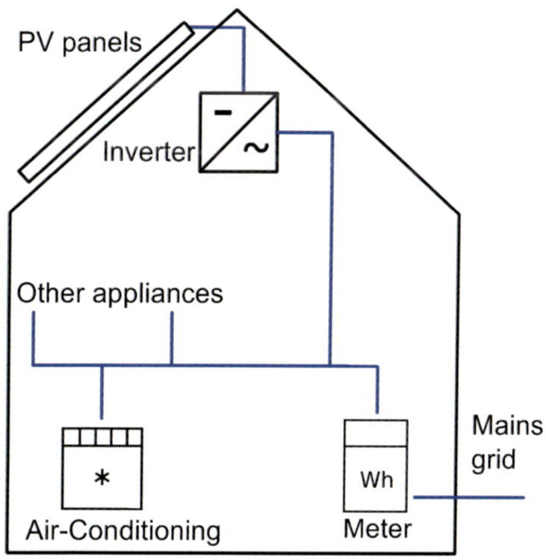

Figure 3.11 Scheme of a standard grid-connected PV system using net-metering. Power generated from the sun is used to reduce the building power consumption, including the consumption for air-conditioning. The difference between solar power generation and building consumption is accounted for in the meter.

Source: Solem Consulting

Some countries have already introduced time-of-use metering, i.e. the metering of power at different electricity prices used at different times of the day. Special electricity meters are required for this. Typically, the highest electricity prices are charged during the middle of the day (peak tariff), while early morning, late evening (shoulder tariff) and night-time consumption (off-peak tariff) comes at a lower cost. PV power generation is typically at peak during midday and can therefore reduce the cost for air-conditioning significantly if such a tariff structure is present. Net-metering conditions will differ from country to country and it is essential to know the details of the local scheme before an assessment can be made of the viability of any potential PV-driven air-conditioning installation.

FiT situations

The situation changes if a FiT subsidy scheme is present. In its essentials the system is the same as in the net-metering situation, except for the position and possibly the type of meters. In a FiT scheme PV power generators are paid a cost-based price for the electricity they generate. Typically, this price is higher than the cost for electricity purchased and fixed for a number of years. In this case the PV power is not used to reduce the building's power consumption. Instead, all power generated is exported to the grid and accounted for. This requires a second meter installed in the building (Figure 3.12). Power generated from the sun is completely exported to the grid. A second meter is used for accounting purposes. The air-conditioning unit is operated by power purchased from the grid.

The power exported to the grid is not used for air-conditioning purposes, so strictly speaking this kind of system might not be considered a solar cooling system as power purchased from the grid is used to operate the air-conditioning unit. But in fact, technically it is exactly the same as the net-metered system, except for the metering arrangement.

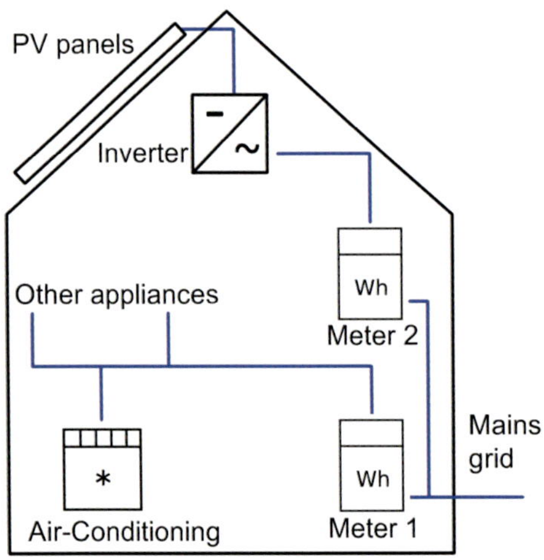

Figure 3.12 Scheme of a grid-connected PV system with a feed-in tariff subsidy scheme.

Source: Solem Consulting

3.5.3 Grid-independent PV cooling systems

Grid-independent PV (off-grid) systems operate without grid connection and require a battery storage to buffer energy for periods with low insolation. A grid-independent PV cooling system can provide good performance in countries where air-conditioning is required most of the day and at night, e.g. the tropical zone (between the tropics of Cancer and Capricorn). Batteries store additional electricity during the day for evening or night-time operation. The battery storage is very expensive and only rarely done. Also, the lifetime of batteries is short and batteries will need to be replaced a few times during the lifetime of the PV panels. Figure 3.13 shows the system schematic with battery storage.

Note: The system in Figure 3.13 can also be extended to supply PV power to other appliances in the building. In this case, PV array and battery size need to be increased accordingly.

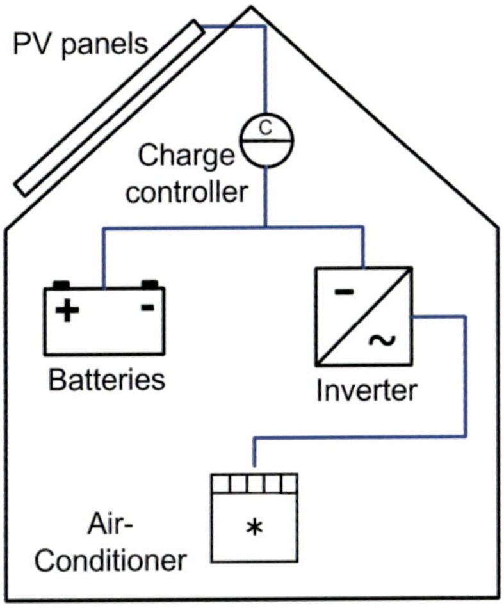

Figure 3.13 System schematic of a grid-independent PV cooling system with battery storage. The charge controller directs the DC power from the PV modules either to the batteries or to the inverter. When no solar power is available, the inverter can operate on battery power.

Source: Solem Consulting

3.5.4 General conclusions about cooling with solar electricity

A definite advantage of solar cooling in grid-connected PV systems is that conventional air-conditioning units are commercially available globally at much lower cost than thermally driven units. The installation of conventional electrical units requires less effort than for thermal-driven air-conditioners, mainly because laying wires is easier (and cheaper) than laying hydronic piping. Apart from air-conditioning, PV power can also be used to reduce the building power consumption at times when the air-conditioning is not needed.

PV cooling (as part of standard grid-connected PV systems) is generally more economically feasible at small-scale capacities (below $30\,kW_r$) than solar thermal cooling. This is mainly due to the difference in equipment and installation cost for these systems. At larger system sizes the situation can be different and in favour of solar thermal cooling. However, this is building-dependent and needs to be evaluated individually for each project. If a building requires heat (e.g. for domestic water or space heating) the situation may also change towards solar thermal cooling.

Away from the electricity grid, powering air-conditioning with PV is a problem because of the need of batteries to store power for when solar radiation levels are low and at night. Batteries are expensive and need to be replaced several times during the life of the system. Therefore, this type of system is extremely rare.

4

Cold distribution

> **Essential terms for cold distribution of solar cooling systems**
>
> **Air-based systems –** open cold distribution systems using air as the heat transfer medium in air ducts to supply conditioned air to rooms and to collect the exhaust air from the building
>
> **Dehumidification –** dehumidification of air through the dew point
>
> **Dew point temperature** – the dew point is the temperature below which the water vapour in air at constant air pressure is condensed into water in the same amount as was evaporated
>
> **Water-based systems** – closed cold distribution systems using chilled water as heat transfer fluid in pipes. Heat is taken up from the space to be cooled by means of different water-to-air heat exchangers

Cold distribution is a general term describing how cold, typically in the form of chilled water or air, is distributed within a building. In other words, it describes the way heat is removed from a building. Cold distribution for solar cooling systems is in principle no different to that of standard air-conditioning or HVAC systems; however, some care has to be taken in the sizing. Different temperature levels (lower temperatures for dehumidification or higher temperatures for sensible cooling only) as well as different heat transfer media (water or air) are possible.

In general, a solar cooling system can operate with both air- and water-based cold distribution. However, the high specific heat capacity of water allows smaller pipe diameters and thus lower investment costs against air-based systems that use larger cross-sections for ducts. Therefore, different options will be presented and discussed to give an overview of water-based systems as well as air-based systems. A cold distribution system has to be designed to meet the required comfort level and architectural aspects. Furthermore, the integration of cold distribution systems in new or existing buildings can pose some challenges, e.g. for low ceiling heights in existing buildings or if no suspended ceilings are desired in office buildings. These issues will be discussed below.

4.1 General design aspects

Before starting with the selection of a cold distribution system the following factors should be clarified and taken into account:

- Is a chilled-water or chilled-air system required? This depends on the building type (new/existing) and comfort requirements (e.g. no ventilation allowed or air exchange mandatory).
- Which chilled water or air supply temperature level is required? This depends on the application and influences the performance of solar cooling systems significantly.
- Is dehumidification of the air supply required? Typically this depends on the building type and location (e.g. high outside humidity).
- For retrofits: is there enough space available for the installation of a water-based or air-based system (e.g. diameter of piping or ventilation duct)?

Depending on the answers to these questions, different cold distribution systems can be selected. These are described below.

4.2 Water-based systems

Water-based distribution systems are closed systems using chilled water as heat transfer fluid in pipes. Heat is taken up from the space to be cooled by means of different water-to-air heat exchangers. Water-based distribution systems (with or without a cold-water storage tank) are typically the same for solar cooling systems and conventional systems. The standard temperature difference is 6/12 °C (43/54 °F) supply/return temperature for air-conditioning applications. The temperature difference between inlet and outlet (ΔT) in the chilled water circuit is usually 3–6 K. A higher temperature level (e.g. 15/18 °C or 59/64 °F) works in favour for any AB- or ADsorption chiller. It leads to an increase of the COP and the available cooling capacity, as discussed in Chapter 6 (see Figure 6.7). However, the temperature level depends on the type of water–air–heat exchanger and comfort requirements of the building (discussed in Chapters 4.2.1–4.2.4).

All cold-water distribution pipework should be insulated to minimise heat gains from the environment. Pipework should not be routed through hot ducts or run adjacent to heat sources such as radiators. Where hot and cold water pipes are run horizontally together, the cold water pipe should be located beneath the hot water pipe to minimise local warming by means of convection.

4.2.1 Fan coils

A fan coil is a simple component consisting of a water-to-air heat exchanger (cooling coil), a filter and a fan. Hot air from the building is blown (or drawn) over the heat exchanger coil and through the filter. It is chilled and enters the building through the coil outlet (see Figure 4.1). Fan coils are the most popular method in water-to-air cold distribution systems and are used to cool or heat the building air in many different types of locations, e.g. residential buildings, offices, supermarkets, warehouses, production areas, etc.

There are two types of fan coils available:

1. fan coils with two pipe connections (one supply and one return pipe) connected to the cold distribution system; and
2. fan coils with four pipe connections (two supply and two return pipes).

The four-pipe fan coil allows both cold and hot water supply at any time, which is often necessary to cool and heat different parts of a building at the same time, e.g. cooling at the east side where the sun heats the office rooms and heating at the still-shaded west side of a building during the morning. Usually a fan coil is used to control the temperature of a single room connected to a central chilled water supply (e.g. from an AB- or ADsorption chiller). The temperature of the room is controlled either by a manual on/off switch or by a thermostat.

Building air dehumidification is achieved if the chilled water supply temperature is lower than the dew point temperature of the air in the building. Then, the low fan coil temperatures will condensate the moisture contained in the hot

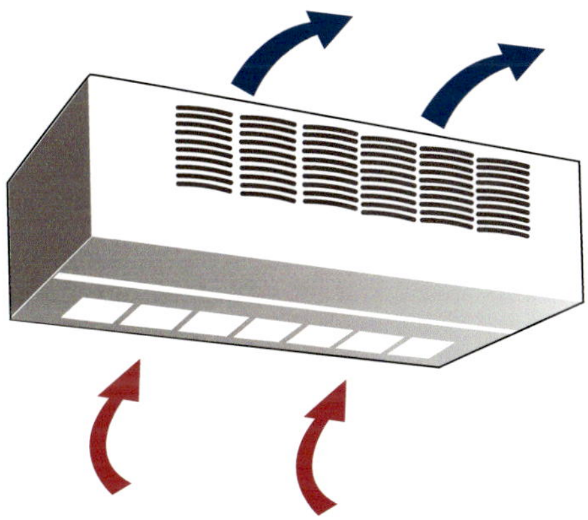

Figure 4.1 Ceiling mounted fan coil with hot air inlet at the bottom and chilled air outlet at the top.

Source: Grundfos

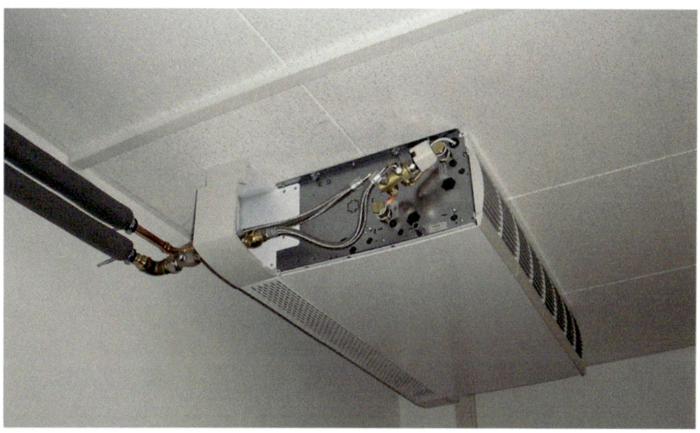

Figure 4.2 Example of two-pipe fan coils with condensate drain connection and chilled water temperatures of 12/6 °C (53.6/42.8 °F) (left) and without condensate drain connection for chilled water temperatures of 18/15 °C (64.4/59 °F) (right).

Source: Solem Consulting (left), SolarNext (right)

air, thus dehumidifying and chilling the air. The condensate water needs to be drained into a drip tray with connection to a pipework laid to fall to a suitable downpipe (see Figure 4.2). If gravity-based draining is not possible then a condensate pump has to be used. Common chilled water temperatures for fan coils are 12/6 °C (53.6/42.8 °F), 13/7 °C (55.4/44.6 °F), 15/10 °C (59/50 °F) or 18/15 °C (64.4/59 °F). Fan coil cooling capacities range between 1 and 12 kW_r per fan coil unit. This cooling capacity can be realised with volume flows of 150 to 1,800 m^3/h, respectively.

Usually fan coils are installed either vertically on the floor (floor mounted) or horizontal on the wall (ceiling mounted). However, fan coils can be noisy because the fan is within the same room and the comfort is not the best due to the air ventilation. But fan coils are simple and more economic to install than any other water-based central cooling systems.

4.2.2 Ceiling panels

An alternative to fan coils – and their comfort restriction due to active air ventilation – are cooled ceiling panels. Ceiling panels are often used as suspended ceilings in office buildings, hotels, etc. They usually have a closed or partially closed surface and the heat transfer is predominantly by radiation (at least 60 per cent). The required space for installation is usually not larger than that for the construction of a suspended ceiling without cooling. In general, there are three different closed cooling ceilings available:

1. suspended closed metal panels with copper or polypropylene tube registers and heat conducting plates (Figure 4.3) or capillary tube mats;
2. suspended plasterboard panels with copper or polypropylene tube register and heat conducting plates (Figure 4.4) or capillary tube mats;
3. plastered ceiling panels with polypropylene capillary tube mats.

Another option are open ceiling panels, where mainly the convective part (at least 60 per cent) is used in the heat exchange. This suspended metal ceiling is an open structure where the convection, and thus the cooling capacity, is increased (Figure 4.5).

The advantage of ceiling panels is the radiation heat transfer and its effect on the room temperature. The room temperature perceived by a person depends on the ambient air temperature and the temperature of the walls, floor and ceiling that surround a person. Heat can be removed from the room by either lowering the ambient air temperature or by lowering the temperature of the surrounding walls, floor and ceiling. The latter is based on the radiation exchange between a person and the surrounding surfaces. It creates a comfortable indoor climate since no cold air movements are present in the room. To achieve radiation cooling, large surfaces in the room must be cooled. Typically, these are the ceiling or the walls. A thermostat including a humidity sensor controls the temperature of the room. Required chilled water temperatures for ceiling panels are 18/15 °C (64.4/59 °F) or 19/16 °C (66.2/60.8 °F) due to the dew point temperature restriction (otherwise condensate water may form and drip from the ceiling). Closed ceiling panels achieve specific cooling

Figure 4.3 Metal ceiling panel with copper tube registers and heat conducting plates made of aluminium.

Source: durlum GmbH, Fotostudio Scheuermann

Figure 4.4 Plasterboard ceiling panels with polypropylene tube register and heat conducting plates made of aluminium.

Source: emco Klimatechnik, Lingen, Germany

Figure 4.5 Open metal ceiling panel with copper tube registers.

Source: www.Lindner-Group.com

capacities between 50 and 100 W/m^2 for suspended metal panels, 60–75 W/m^2 for suspended plasterboard panels, 60–85 W/m^2 for plastered cooling panels and up to 150 W/m^2 for open metal panels.

Single panels

Another option are single open ceiling panels. These are individual elements which are installed as a suspended section of the ceilings. Contrary to closed panels they do not cover the whole ceiling area (see Figure 4.6).

Figure 4.6 Open ceiling panels as single elements installed at an office space.

Source: Carpus+Partner AG, Aachen, Krantz Komponenten

The remaining ceiling areas can be supplemented with a suspended ceiling system, or they may remain visible (Figure 4.6). The difference between closed ceiling panels and single ceiling panels is mainly the higher cooling capacity per unit, which can be achieved with a single panel. The reason is that the open sides of a single panel allow airflow, wherein the air above the cooling panel is also cooled before it falls into the room again. Due to the higher convection, single panels have a higher specific cooling capacity than closed ceiling panels. Note: it is important to design the corresponding distances and suspension height properly to provide convection.

Again, a thermostat including a humidity sensor controls the temperature of the room. In general, there are two different types of single panels with different shapes (flat, triangular, concave or convex, undulating canopy, etc.) available:

1. suspended metal cooling panels with copper or polypropylene tube registers (Figure 4.6);
2. suspended plasterboard cooling panels with copper or polypropylene tube register.

Another option are baffle cooling panels, where single lamellar modules with variable centre distances are suspended from the ceiling (see Figure 4.7). The lamellar modules are made of steel or aluminium and include a heat-conducting plate.

Usually required nominal chilled water temperatures for cooling panels are also 18/15 °C (64.4/59 °F) or 19/16 °C (66.2/60.8 °F) due to the dew point temperature. Single panels achieve specific cooling capacities of 80 and 130 W/m^2 for metal cooling panels, 80–100 W/m^2 for plasterboard cooling panels and up to 250 W/m^2 for plate finned cooling panels. The combination of concrete core activation (see Chapter 4.2.4) with ceiling panels is also an option, which is useful if very high cooling loads exist, such as in a fully glazed office building with high external as well as internal loads.

Figure 4.7 Open constructed chilled baffle cooling panels in an office space.

Source: www.Lindner-Group.com

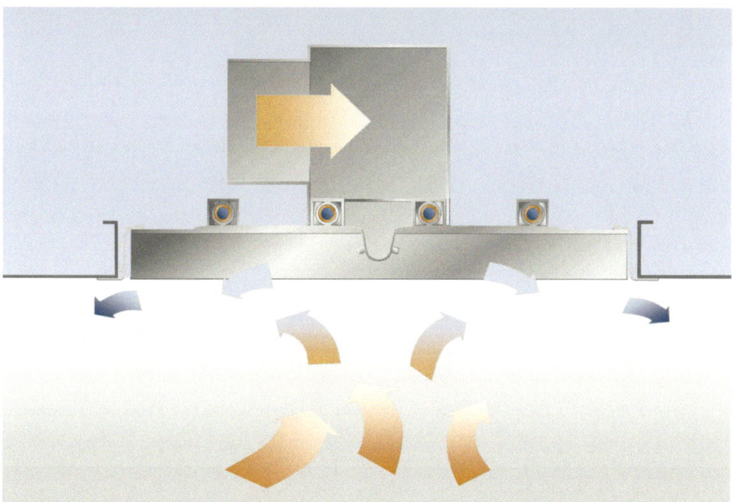

Figure 4.8 Cooling panel with integral air diffuser.

Source: Kiefer GmbH

Ceiling panels can also be combined with a fresh outside air supply; for example, for production areas. These panels have an integral air diffuser (see Figure 4.8), which improves the heat transfer of the panels by using outside air and increases the specific cooling capacity up to 250 W/m^2.

Another approach with ceiling panels is the integration of micro-capsuled PCM (phase change material), such as paraffin wax, into the plasterboard with two exact definable switching temperatures, at either 23 °C (73.4 °F) or 26 °C (78.8 °F). Such panels can absorb the heat loads from rooms by using the latent heat storage properties of the paraffin wax, keeping the room temperature constant until the total storage capacity of the paraffin wax is exhausted. Only then is the active cooling system, which is connected to polypropylene capillary tube mats, switched on. The specific cooling capacity of these cooling panels is about 70 W/m^2.

Figure 4.9 Plasterboard cooling panels with integrated PCM.

Source: Ilkazell Isoliertechnik GmbH

4.2.3 Concrete core activation

Concrete core activation can be an economical method for cooling of buildings. It involves utilizing the ceiling mass of a building to store thermal energy and thus cool rooms. This system is particularly suitable for offices, administrative and school buildings, or modern glass architecture with concrete floors. Prefabricated plastic piping systems are installed within the reinforcement layers in the concrete components of the building (see Figure 4.10). Water is circulated in the pipes, the rate of circulation determined by the temperature of the ceiling. The nominal chilled water temperatures for concrete core activation is usually between 19/16 °C (66.2/60.8 °F) and 21/18 °C (69.8/64.4 °F). Instead of chilled water supplied by an active cooling system, the concrete core activation could also be combined with free cooling (see Chapter 6) to save energy. Instead of water, outside air can also be used as a heat transfer medium, but the required cross-sections of the ducts are larger, making their integration into the concrete building components more difficult.

The closer the plastic pipes are arranged on the surface, the greater the achievable performance. With pipes that are fixed in the middle of the static neutral zone of the concrete ceiling, specific cooling capacities of about 35 W/m^2 can be achieved. This is often only enough to cover the base load. But if the pipes are fixed onto the concrete surface, specific cooling capacities of 90 W/m^2 can be reached. A thermostat, ideally in combination with a humidity sensor, controls the temperature of the room.

The performance of a concrete core activation system is reduced by suspended ceilings, hence there are special ceiling panels that operate on the basis of concrete core activation and produce only a small cooling capacity loss (see Chapter 4.2.2). The operation of a concrete core activation system can also be improved by operating the system during night-time to dissipate the stored heat from the concrete ceiling in order that the concrete can absorb more heat again the next day.

Figure 4.10 Plastic pipes in the reinforcement of a concrete core activation.

Source: Zewotherm

4.2.4 Floor/wall cooling

Conventional installed underfloor or wall-heating systems can also be used in summer as floor or wall cooling systems. These can be either surface cooling (pure water system) or convectors (an air/water system with special office furniture incorporating chute or basic floor convectors). The nominal chilled water temperatures for floor cooling are usually between 20/17 °C (68/62.6 °F) and 22/19 °C (71.6/66.2 °F). Wall cooling systems can operate down to 18/15 °C (64.4/59 °F). The temperature of the room is controlled by a thermostat. In general, there are two different floor/wall surface cooling systems available:

1. wet systems with copper or polypropylene tubes or capillary tube mats;
2. dry system with copper or polypropylene tubes.

Wet systems are available for floor cooling, in which the pipes are installed in the screed (very often cement screed or anhydrite due to the better heat transfer in liquid screed (see Figures 4.11 and 4.12). In dry systems the pipes are located below the floor covering. The pipes are fixed on a carrier insulation, which is provided with heat-conducting strips (Figure 4.13). The strips provide for better cold distribution. Dry systems are especially suitable for low floor constructions and are used in old houses or during building renovation/modernisation.

Wall cooling systems also come as either wet or dry systems. These are basically the same as floor cooling systems, except that they are integrated into walls rather than floors. So-called wet systems are primarily used in solid construction, in which a reinforcement of metal, mineral fibres or plastic fibres is integrated into the wall plaster. Dry wall cooling systems are also commercially available, in which either heat-conducting aluminium strips are fixed to plasterboard, or the tubes are inserted directly into the plasterboard. These systems are already partially available as prefabricated panels.

With floor cooling, specific cooling capacities between 30 and 65 W/m^2 can be achieved. Against that, only 20–50 W/m^2 can be reached with wall cooling systems. Dry systems generally have a lower specific cooling capacity.

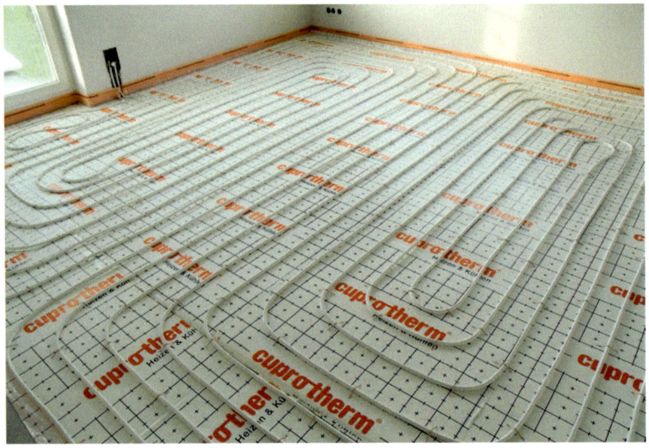

Figure 4.11 Wet floor cooling system with copper tubes.

Source: Wieland

Figure 4.12 Wet floor cooling system with capillary mats.

Source: BEKA

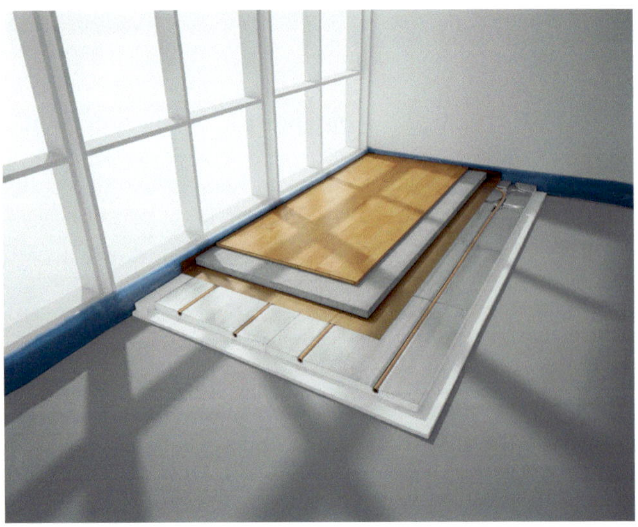

Figure 4.13 Dry floor cooling with copper tubes.

Source: JOCO Wärme in Form, Willstätt

4.3 Air-based systems

Air-based distribution systems are open systems using air as the heat transfer medium in air ducts to supply conditioned air to the rooms and to collect the exhaust air from the building. Air distribution is required for solar cooling systems using solid or liquid DEC (see Chapter 3.2). Function and sizing is the same as for conventional air-handling units. Thus air-based systems are only a one-dimensional problem – always the same kind of air transport and similar air diffusers, regardless of temperature level, unlike water-based systems. The temperature of the air supplied to the building interior is usually between 16 °C and 20 °C (60.8–68 °F). Air-based systems are used in office buildings, shopping malls, libraries, hospitals and airports, where air volume flows are building-specific. For example, an air volume flow of 10,000 m^3/h corresponds to the fresh air requirements of about 300 people in an office building. The distribution of the air supply is typically via air ducts along a corridor below the ceiling. These ducts branch off into the individual rooms, where air diffusers are used to distribute the air in the room itself. Exhaust air is removed via air ducts from rooms like kitchens, bathrooms or toilets. The corridors between serve as overflow zones (e.g. when doors are left open).

Note: the pressure losses in the system (filter, heat exchanger, etc.) and the air duct distribution system need to be kept low; otherwise power requirements for the fans can be significantly increased.

Sometimes air/water heat exchangers are used to additionally cool down the air supply before it is pumped into a room (see Figure 4.14). This configuration is beneficial if humidity levels are high and the DEC system has limited dehumidification capability (due to its size) with regard to the outside air drawn into the building.

Figure 4.14 Air supply duct with air/water heat exchanger for cooling purposes.

Source: SolarNext

4.4 Cold distribution: system comparison and overview

An overview of water-based and air-based cold distribution systems is provided in Table 4.1. Figure 4.15 shows the different available water-based systems and their corresponding chilled water temperature/spread range. Note: high efficiencies of AB- or ADsorption chillers are possible with high chilled water temperatures above 15 °C (59 °F). Fan coil applications for 15/10 °C (59/50 °F) temperature spread are also possible; standard spread of 12/6 °C (53.6/42.8 °F) is not always necessary.

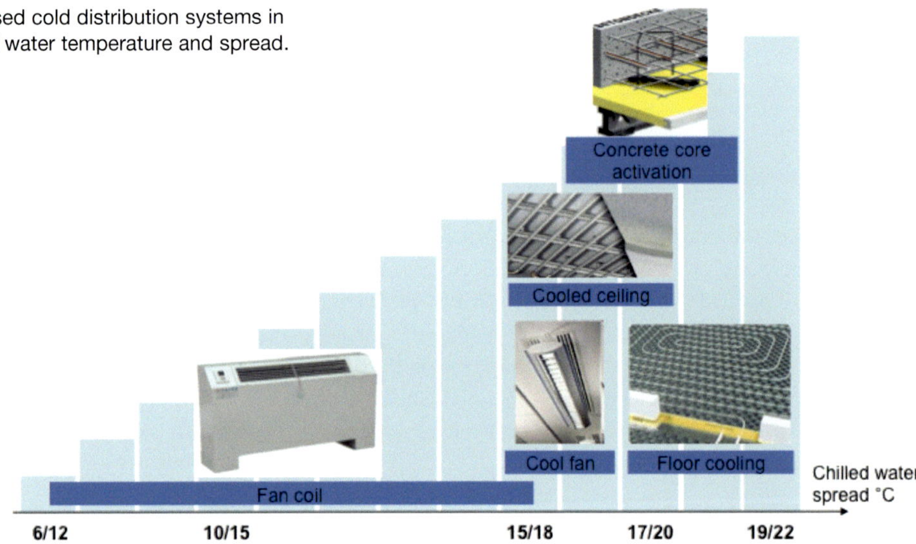

Figure 4.15 Water-based cold distribution systems in response to the chilled water temperature and spread.

Source: SorTech

Table 4.1 Summary showing characteristics of water- and air-based cold distribution systems

	Fan coil	Closed ceiling panels	Single ceiling panels	Concrete core activation	Floor/wall cooling	Air duct
Specific cooling capacity	1–12 kW_r per unit	50–100 W/m² (150 W/m²)[a]	80–130 W/m² (250 W/m²)[b]	35–90 W/m²	20–65 W/m²	Adiabatic cooling
Temperature	6–15 °C (42.8–59 °F)	15–16 °C (59–60.8 °F)	15–16 °C (59–60.8 °F)	16–18 °C (60.8–64.4 °F)	17–19 °C (62.6–66.2 °F)	16–20 °C (60.8–68 °F)
Temperature spread	3–6 K	3 K	3 K	3 K	3 K	–
Dehumidification possible	Yes	No	No	No	No	Yes
Fan power required	Yes	No	No	No	No	Yes
Installation as retrofit possible	Yes	Yes	Yes	No	No	Yes

Notes

[a] Specific cooling capacity applies to open metal ceiling panels only.

[b] Specific cooling capacity applies to plate finned and cooling panel with integral air diffuser only.

5

Storage in solar thermal cooling systems

Essential terms for heat and cold storage

Aquifer – a layer of permeable rock, gravel, sand or silt that contains groundwater. Extraction of the water is usually via boreholes

Density – the mass unit per volume of a substance

District heating/cooling – a process where heat or cold is distributed from a central plant via insulated pipes to residential and commercial requirements

Latent heat – latent heat refers to the energy change of a substance when changing phases at constant temperature (latent heat of fusion for melting and freezing; latent heat of vaporisation for boiling and condensing)

Load profile – the distribution of heating and cooling requirements of a building over time. The load profile can be expressed for daily, weekly, monthly and annual time periods

PCM – phase change material; a material that is used in latent heat storage applications. PCMs are chosen according to the temperature level of the storage application

Sensible heat – sensible heat refers to the energy change of a substance that has a change of temperature as its only cause (no phase change)

Solar fraction – the fraction of the total annual hot water, heating or cooling requirements of a building supplied by a solar system

Solar yield – the useful thermal energy from solar thermal collectors; usually calculated per year

Specific heat capacity – the quantity of heat required to raise the temperature of a given mass of substance by a given amount

Thermal conductivity – the ability of a substance to conduct heat

Thermal (ground) probe – a U-shaped pipe that is inserted into sealed vertical or inclined borings. Water is pumped through the pipe, which then acts as a heat exchanger and either absorbs the heat from the earth or releases heat into the earth.

Thermal storage of heat and/or cold is an important component of a solar cooling system. This chapter gives information on how heat and cold can be stored. The correct design of storage components is a must for sound and reliable system operation and performance. It is important to differentiate between the heat storage component of a system (heat stored to drive the system) and the cold storage component (cold air or water to be distributed for space cooling purposes). Although storage is not generally necessary for overall system operation, it does provide some important advantages:

1. *Decoupling of solar yield from load profile.* During periods of high solar insolation and low cold demand, additional heat can be stored and used later. Alternatively, additional cold can be generated during this period and also used later (see Figure 5.1).
2. *Evening or night-time operation without sun.* Heat or cold stored during the day can be used in periods without sunshine.
3. *Cloud buffering.* Periods of intermittent cloud cover during the day can be bridged in order to maintain continuous system operation.
4. *Robust system control.* Potential temperature fluctuations caused by changing ambient conditions (sun, wind, humidity) can be balanced out by the storage tank. Less control action is required to maintain a stable cold supply temperature.
5. *Provision of heat for other purposes.* If hot water and/or space heating is required in addition to the cooling, a storage tank provides means to supply these. This is especially important in locations with winter heat demand.

The requirements of a thermal storage system from a technical and economic point of view are that it:

- has a low volume and low weight;
- has low thermal losses;
- is low cost.

Storage technologies can be classified according to construction, active principle and applicable temperature range. An overview is given in Figure 5.2.

Low volume and weight requirements can be translated into two physical parameters of the storage material: density and specific heat capacity. The more dense a material and the more heat it can store, the less volume and weight is required for a given storage requirement. For some storage materials the heat conductivity is important, e.g. when solids are being used to store heat. The heat conduction through a solid material determines the rate at which a storage tank can be charged/discharged. Figure 5.3 shows an overview of common thermal storage materials and their main physical parameters.

Thermal loss can be minimised by sufficient insulation of the storage tank, but also by having a low surface area–volume ratio of the tank. The difference in surface area–volume ratio for small and large cylindrical tanks is shown in Table 5.1. It can be seen that larger tanks have less surface per volume, thus reducing thermal losses (via the surface of the tank), incurred through conduction and convection to the external environment.

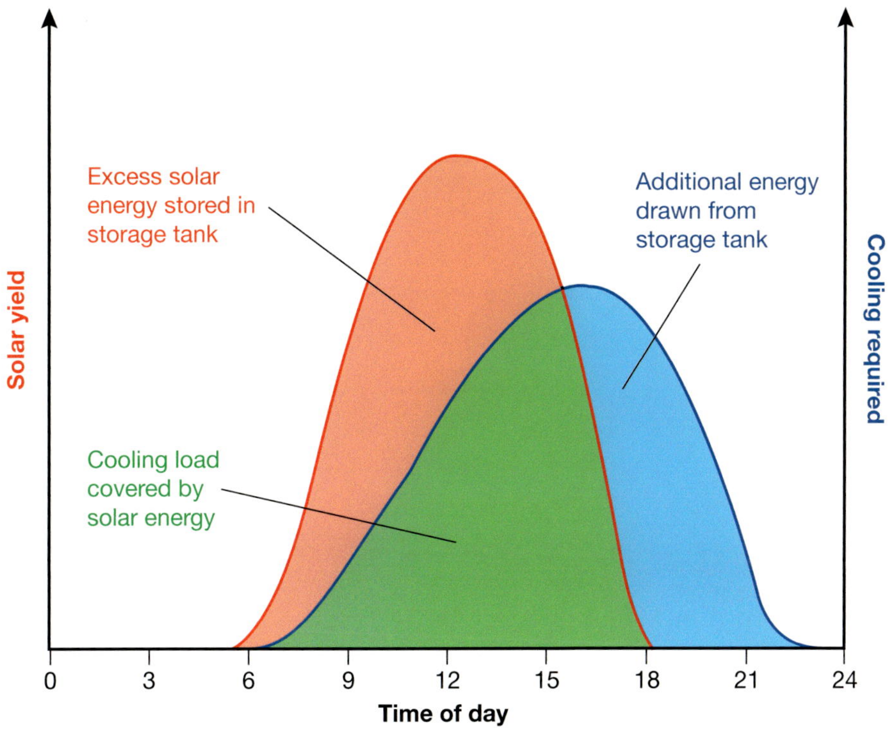

Figure 5.1 Example of decoupling solar yield (orange) from load profile (blue). The green hatched area symbolises the fraction of cooling required that is covered by solar energy. While the cooling load is still low, excess solar energy can be stored in a tank (orange hatched area). Later in the day, the load exceeds the solar yield and additional energy is drawn from the tank (blue hatched area).

Table 5.1 Surface-to-volume ratios of various cylindrical tanks

Tank diameter (m)	**Tank height (m)**	**Tank volume (m^3)**	**Tank surface[1] (m^2)**	**Surface-to-volume ratio (m^2/m^3)**
0.5	2	0.39	3.5	9.0
2	0.5	1.57	9.4	6.0
1	2	1.57	7.9	5.0
2	1	3.14	12.6	4.0
4	10	125.6	150.8	1.2
10	4	314.2	282.7	0.9

Note:

[1] Tank surface has been calculated including the bottom of the cylinder (its footprint)

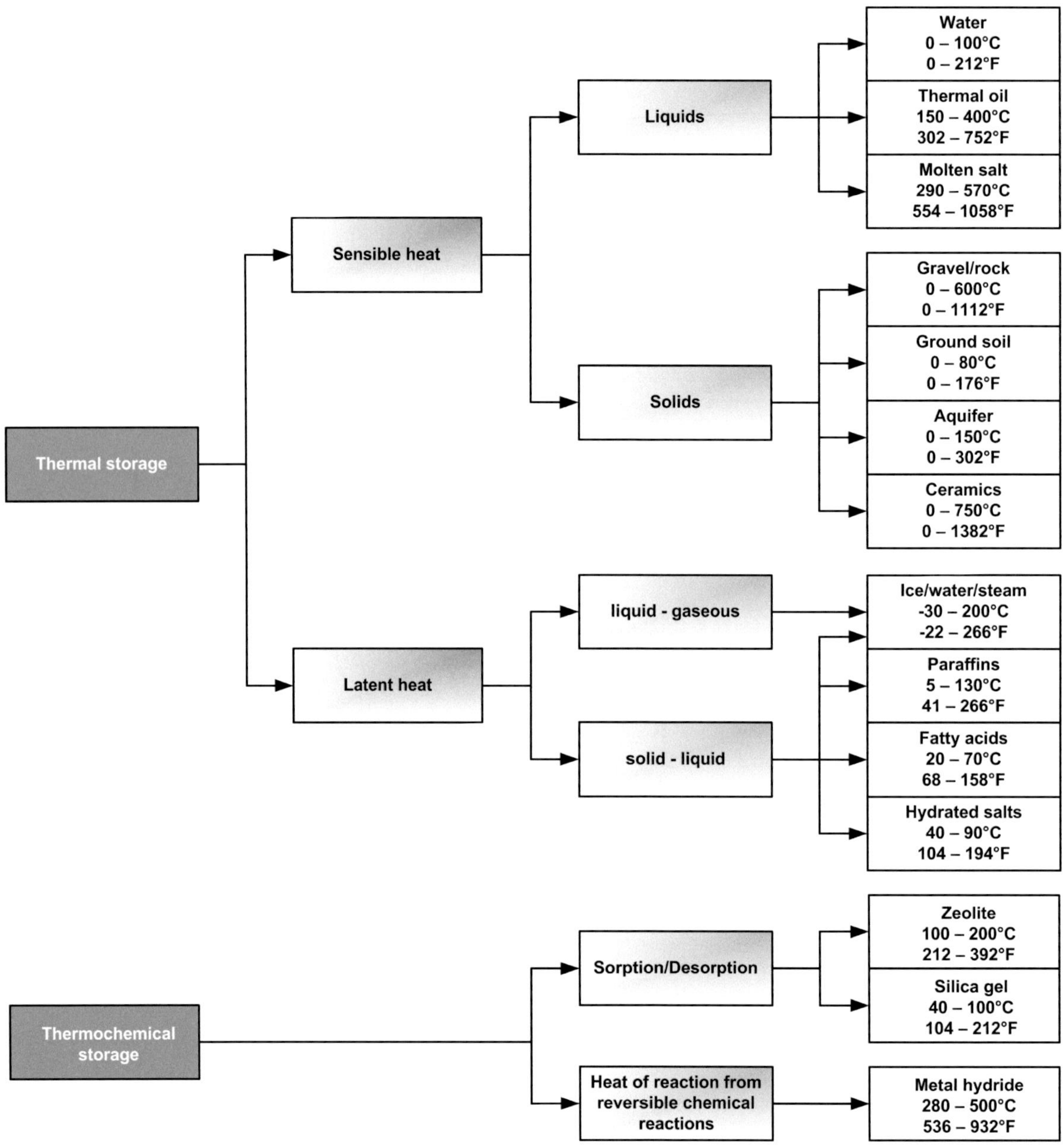

Figure 5.2 Overview of thermal and thermochemical heat storage technologies. Thermal storage options include both heat and cold storage, thermochemical storage options include heat storage only. Example: Heat at 90 °C (194 °F) can be stored as sensible heat (without a phase change) using one of the following options: liquid water, gravel/rock, aquifers or ceramics. If a phase change is to be used (latent heat) then the following options are possible: paraffins or hydrated salts. The use of thermochemical storage zeolites and silica gel are also possible.

Source: Solem Consulting

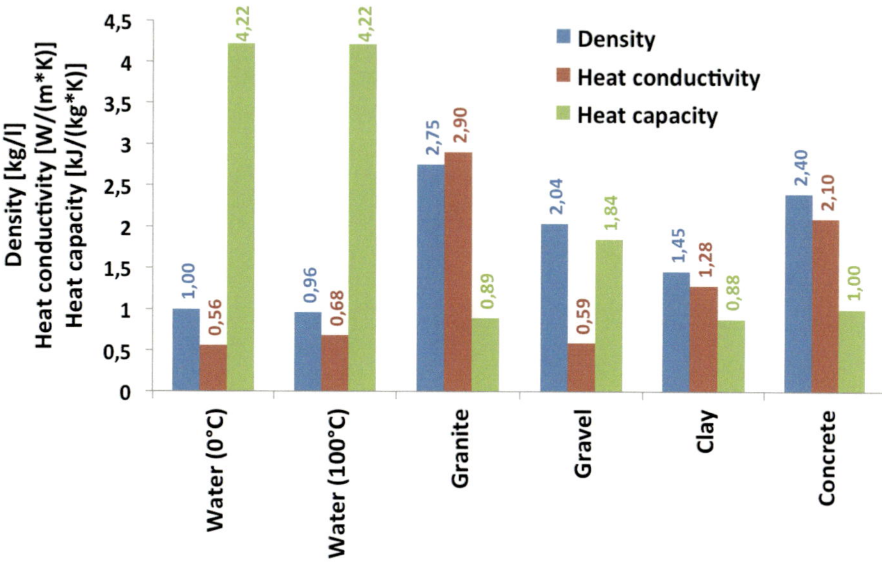

Figure 5.3 Overview of thermal storage materials and their main physical parameters.

Source: [8]

It is important to differentiate between the heat storage component of a system (heat stored to drive the system) and the cold storage component (cold air or water to be distributed for space cooling purposes). Heat storage is used to store heat from the solar collectors; cold storage is used to store cold from a thermal cooling process. Both differ from each other in the tank type used and their position in the system. Figure 5.4 shows a schematic of the storage options in a closed solar cooling system. The heat storage tank is on the left side of the cold generation, storing heat for either extended operation of the cold generation or for supplying a space heating/hot water load. The cold storage tank is located on the right side of the cold generation and stores chilled water or ice. A thermal chiller can provide chilled water if surplus solar energy is available. Alternatively, cold air can be generated from the stored heat using a DEC system. In this case, storage of cold air is not possible.

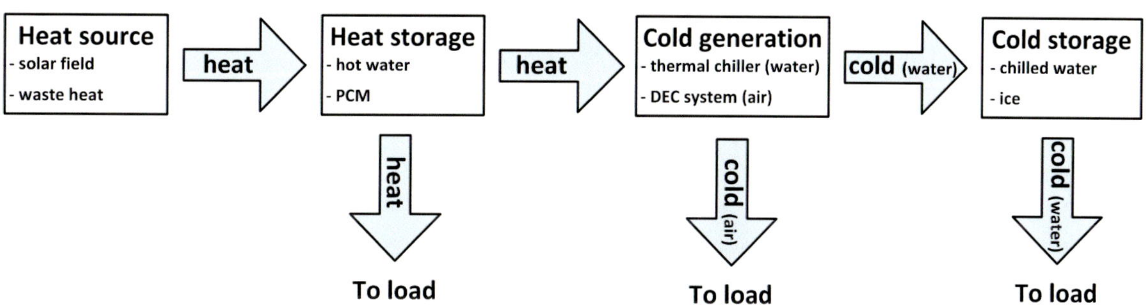

Figure 5.4 Schematic of the storage options in an example solar cooling system.

Source: Solem Consulting

5.1 Heat storage

The most common storage material for heat is water. It is readily available, cheap, non-toxic and environmentally benign. Figure 5.5 shows that water also has a high specific heat capacity compared to other materials. Water-based heat storage tanks are available in many variations for different applications. A difference can be made between water storage tanks for small applications and for larger applications.

5.1.1 Small-scale hot water storage (<10 m^3)

Hot water storage tanks for single family houses and small commercial applications are usually in the range of 300–1,000 litres. Tanks of up to 10,000 litres of hot water volume are used in solar-heated low-energy houses in order to achieve very high solar fractions. Small-scale hot water tanks usually provide storage for a period of a few days to a week. The most common tank types for small applications are shown below.

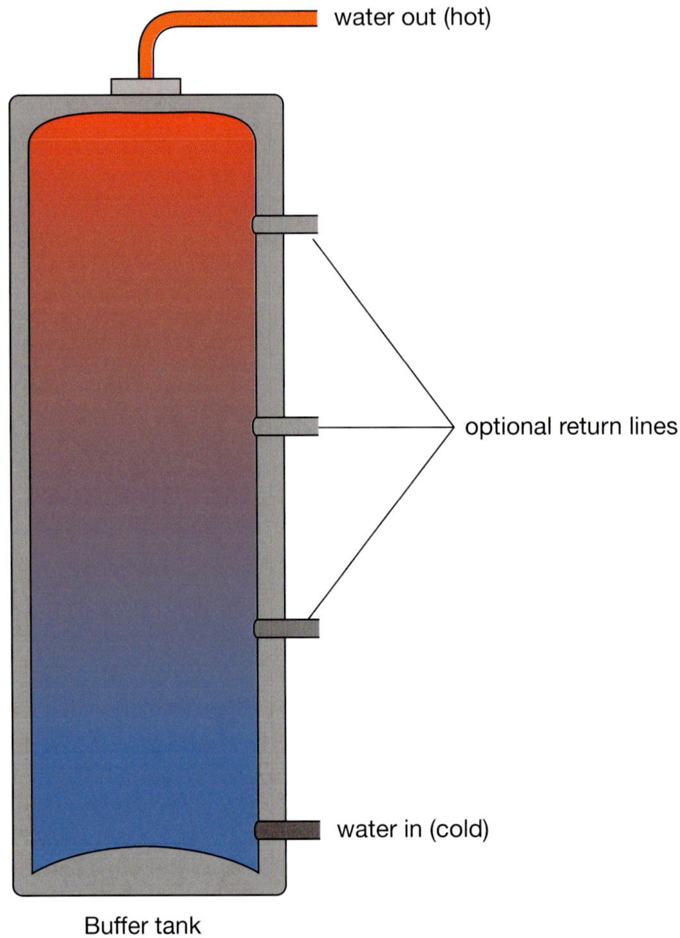

Figure 5.5 Schematic of a simple solar (buffer) tank for non-potable water.

Figure 5.5 shows the schematic of a simple solar tank, also known as a buffer tank, a plain steel, non-enamelled vessel with no internal heat exchangers. It can be used for non-potable water only and provides heat or cold storage in the system. The buffer tank is the most common tank used in solar cooling systems with no potable water requirements. They come in a wide range of sizes and can usually be manufactured or purchased locally. It is exposed to the operating pressure and temperature of the solar system and needs to be designed accordingly.

A solar tank that holds potable water is shown in Figure 5.6. The solar tank is an enamelled or stainless steel vessel with at least two internal heat exchangers. The bottom heat exchanger transfers heat from the solar collectors into the water body, the top one can provide additional heating from a backup non-solar heat source. The tank has a constant influx of fresh, oxygenated cold water, thus a sacrificial anode in the tank is required to provide corrosion protection. The suggested use of this tank is for heat storage only.

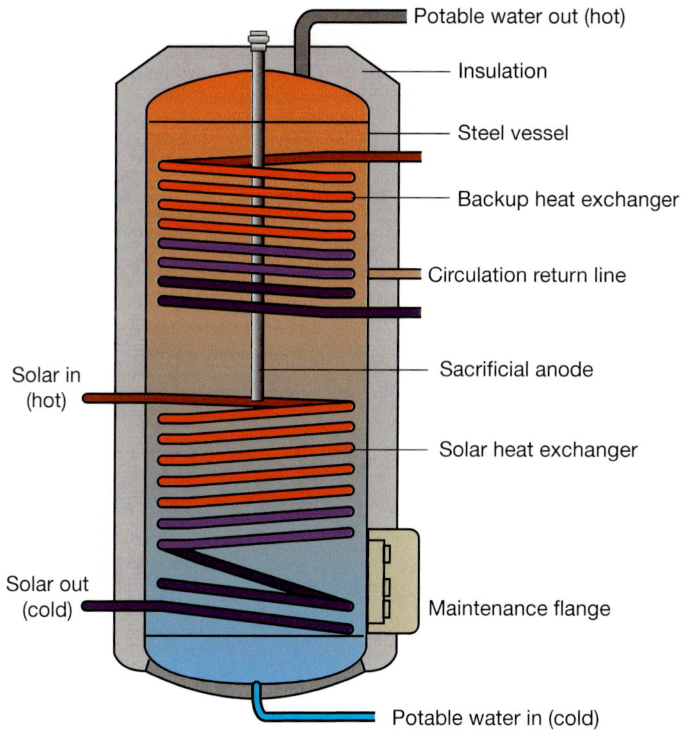

Figure 5.6 Schematic of a solar tank with internal heat exchangers for potable water.

An alternative for hot potable water storage is a so-called combined solar tank, known as a Kombi tank in Europe (see Figure 5.7). It is a tank-in-tank arrangement, with a large external non-potable water tank for the heat storage and a smaller internal potable water tank. The bottom heat exchanger transfers the solar heat into the external water body, which heats the internal potable water tank via heat conduction. Corrosion protection is realised via a sacrificial anode or by using a stainless steel inner tank. This tank is usually used for heat storage only.

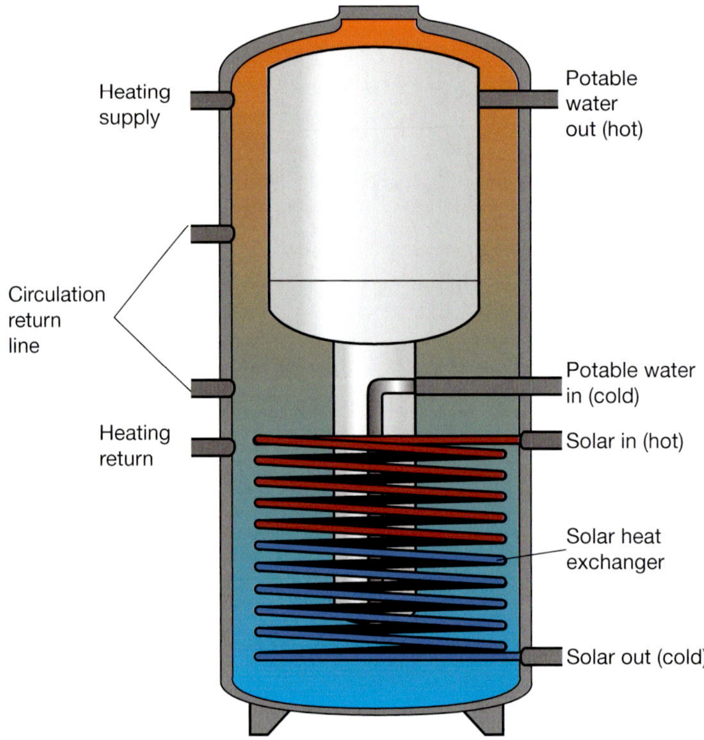

Figure 5.7 Schematic of a combined solar tank for both potable and non-potable (heating) water.

Two other commercially available tank designs are also worth mentioning. The first one is a stratified storage tank, which actively supports the thermal layering of water at different temperatures within the tank (Figure 5.8). A stratification tube is mounted vertically in the tank with various openings at different heights. Hot water rising inside the stratification tube (sometimes also called stratification lance) is being released only into the thermal layer with a matching temperature. Thus, mixing losses within the tank are minimised. The shape of the tank and the water movement inside the tank both influence the stratification. Tall and thin tanks have better stratification than short and squat ones. Using heat exchangers or reducing the speed of pumps will improve stratification as mixing losses between layers are reduced [10].

The other design is a storage tank with an integrated back-up heater (see Figure 5.9). The back-up heater can either be gas- or oil-fired and heats the tank water via a heat exchanger inside the tank. Further, a heat pump option as well as a district heat option are available for this design. In all cases, fewer components are required (no pump, piping, etc.) by integrating the back-up heater directly into the tank. This reduces installation costs.

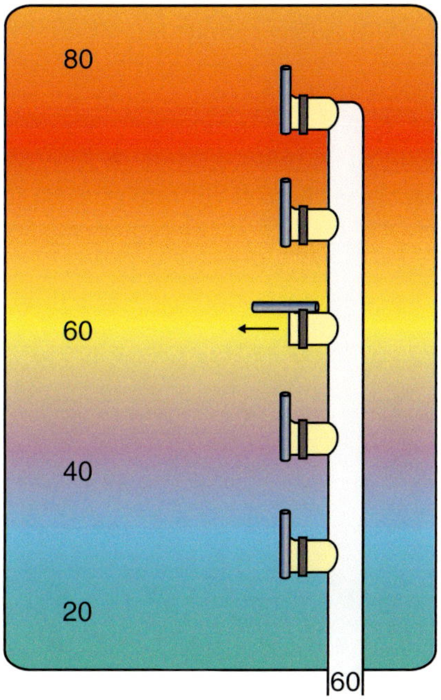

Figure 5.8 Principle of a stratified storage tank. Different temperature layers are present in an unmixed tank. Hot water from the solar collectors at, e.g. 60 °C (140 °F), rises inside the stratification tube and is released into the thermal layer with matching temperature. The stratification tube has various outlets with flaps. These flaps open if the density (and thus temperature) of the water in the tube equals the density of the water on the tank side of the flap.

Figure 5.9 Example of a hot water storage tank with a gas-fired back-up heater integrated into the tank.

Source: Solvis GmbH & Co. KG

5.1.2 Large-scale hot water storage (>10 m^3)

Large-scale heat storage is employed in applications such as commercial buildings or district heating/cooling systems. The higher cost for these systems is usually prohibitive for small applications. Also, large-scale storage is used for longer-term storage, so-called seasonal thermal energy storage (STES). The main principle of STES is to collect heat or cold whenever it is available and to store it for later use. For example, summer heat from solar collectors can be stored and used in winter for space heating. Winter cold can be stored and used in summer for air-conditioning purposes. Large tanks are needed for STES because the quantity of heat or cold to be stored is large and the thermal losses need to be minimised – otherwise the heat/cold stored is lost again over the long time period between seasons.

Large water tanks can be freestanding above ground or buried underground. The state-of-the-art are freestanding, insulated steel vessels, holding some 10,000 m^3 of water. Underground tanks have also been constructed, but in much smaller numbers than freestanding tanks. Several demonstration projects with underground tanks are currently ongoing to investigate the technical and economical feasibility of these tanks. A novelty are hybrid tanks, where an underground water tank is surrounded by thermal ground probes. The thermal loss of the tank is recovered partially by the thermal probes, which are connected to a heat pump (see Figure 5.10).

Sometimes solids are mixed in with the water, such as a water–sand–gravel mixture, to enhance the mechanical stability of large underground tank structures. The surface-to-volume ratio is more favourable for larger tanks, hence the thermal losses are smaller (see Table 5.1). However, the longer pipe runs in district heating/cooling systems can incur high thermal losses. Deciding on a large-scale water tank depends on both thermodynamic and economic calculations.

Figure 5.10 Example of a hybrid water storage tank during construction. The inner tank has a volume of 500 m^3 and is used for hot water storage. Only the roof of the tank is thermally insulated, the side walls are not insulated. Thermal ground probes surrounding the tank recover the heat lost through the tank walls. The tank is used for a solar district heating system in Germany.

Source: M. Reuss

Figure 5.11 shows different underground storage technologies for large-scale hot water storage. Sizing guidelines for these can be given as follows, assuming a solar fraction of 50 per cent for the whole system [9]:

- hot water heat storage, 1.5–2.5 m^3 storage volume per m^2 flat plate collector area;
- gravel–water heat storage, 2.5–4.0 m^3 storage volume per m^2 flat plate collector area;
- geothermal probes heat storage, 8.0–10.0 m^3 storage volume per m^2 flat plate collector area;
- aquifer heat storage, 4.0–6.0 m^3 storage volume per m^2 flat plate collector area.

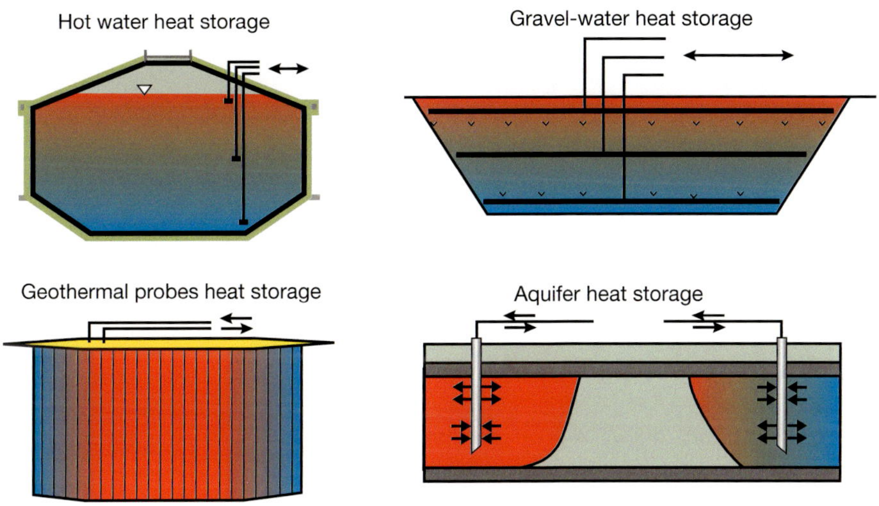

Figure 5.11 Examples of large-scale underground storage technologies.

Source: ITW Stuttgart, Germany

5.1.3 Phase change material

The phase change of materials, typically from solid to liquid and vice versa, is also used to store heat. The latent heat of fusion is the thermal energy absorbed by a material when changing phase at constant temperature. It is much greater than the thermal energy that materials can absorb without a phase change. Common PCMs for heat storage in solar cooling applications are paraffins, fatty acids and hydrated salts. Table 5.2 gives an overview of the relevant physical properties of these. PCMs melt when heat is added to the tank and they solidify when heat is taken out of the tank.

The phase change occurs at constant temperature or within a narrow temperature band. This is advantageous for the solar collector efficiency – the mean collector temperature does not increase during the melting process. In water tanks using sensible heat storage the increasing water temperature in the tank results in rising return temperatures to the collector, thus increasing the collector mean temperature. In turn, the collector efficiency drops (see

Figure 2.23). PCMs in storage tanks are usually doped with nucleating agents to prevent sub-cooling of the material below the melting point.

In comparison to a chilled water tank, PCM storage requires a much smaller volume for the same amount of heat stored. Also, the thermal loss of a PCM tank is smaller due to the uniform temperature distribution in the tank (see example below).

Table 5.2 Physical properties of selected PCMs for heat storage in solar cooling applications.

PCM	Melting point (range) (°C (°F))	Latent heat of fusion (range)[1] (kJ/kg (Btu/lb))
Paraffins	5–130 (41–266)	180–205 (77–88)
Hydrated salts	40–90 (104–194)	230–600 (99–258)
Fatty acids	20–70 (68–158)	130–190 (56–82)

Note:

[1] Latent heat of fusion is the energy change of a substance when changing phases (melting or freezing) at constant temperature.

Heat storage volumes – worked example

A heat storage tank with 100 kWh$_{th}$ (341,442 Btu) of heat to be stored is to be designed using (a) water and (b) a suitable paraffin with a latent heat of fusion of 180 kJ/kg (77 Btu/lb) and a density of 800 kg/m^3 (50 lb/cu.ft). The supply temperature required is 65 °C (149 °F) and the temperature difference between supply and return is 5 °C. Which tank volumes are required for both options?

Answer:
For the sensible heat water tank (no phase change) the required volume can be calculated as:

$$V_{tank} = \frac{Q_{stored}}{c_p \cdot \Delta T \cdot \rho_{water}} \qquad (5.1)$$

where

Q_{stored} is the amount of heat to be stored in the tank [kJ]

c_p is the specific heat capacity of water [kJ/(kgK)]

ΔT is the temperature difference between supply and return of the tank [K]

ρ_{water} is the density of water [kg/m^3]

The water volume V_{tank} required in this case is 17.2 m³ (607 cu.ft).

For storage using PCMs the volume can be calculated as:

$$V_{tank} = \frac{Q_{stored}}{L_f \cdot \rho_{PCM}} \tag{5.2}$$

where

Q_{stored} is the amount of heat to be stored in the tank [kJ]

L_f is the latent heat of fusion of PCM [kJ/kg]

ρ_{PCM} is the density of liquid PCM [kg/m³]

In this case, the required tank volume is 2 m³ (70 cu.ft) or less than one-eighth of the water tank volume. Note that these calculations exclude the thermal losses of the tank and associated equipment. So, in reality, both tanks would have to be larger to account for this.

While the PCM storage tank volume is much smaller in the boxed example, one has to note that the peak thermal power of PCM tanks during discharging (solidifying) is lower than that of water tanks. This is because of the growing layer of solid PCM material around the heat exchangers in the tank during solidification. The solid PCM material has a higher thermal resistance than the liquid material, and once it has formed around the heat exchanger surfaces, less heat can be transferred from the liquid PCM to the heat exchanger. The solid PCM material is like an insulation layer, thus reducing the peak thermal output of the tank.

The constant melting temperature of PCM results in small temperature differences between external supply and return when charging/discharging a PCM tank. This can be a disadvantage, since not all thermal cooling processes are suited to such small temperature differences. Therefore, any combination of PCM tank and thermal chiller/DEC system needs to be calculated carefully with regards to this.

5.2 Cold storage

Cold storage tanks use both sensible and latent energy storage. Chilled water tanks are the most common; however, ice and slurry tanks are also state-of-the-art. Storing cold instead of heat has advantages from a thermodynamic point of view. The temperature difference between a cold storage tank and the ambient is lower than between a heat storage tank and ambient, so energy losses are lower (see box).

Heat loss of a cold tank – worked example

A chilled water tank storing cold at 6 °C (42 °F) has a temperature difference of 19 °C to an ambient temperature of 25 °C (77 °F). A hot water tank storing heat at 75 °C (167 °F) has a temperature difference of 50 °C to ambient. The heat loss of a tank due to conduction through the tank wall depends linearly on the temperature difference, hence the heat storage tank in our example has a thermal loss approx. 2.5 times greater than the cold tank. Note that this example neglects any heat losses due to convection outside the tank.

5.2.1 Chilled water

Chilled water tanks are state-of-the-art in conventional air-conditioning technologies. Usually they are installed to take advantage of electricity day–night price differences. The storage tank is charged at night using conventional electrically powered chillers and lower-price electricity. The stored cold is then used to provide air-conditioning during the day. Tank sizes range from a few thousand litres up to several million litres (see Figure 5.12).

The use of chilled water tanks in solar air-conditioning systems is determined by the usable temperature difference of the chilled water load, i.e. the difference between supply and return temperatures. The standard temperature difference for chilled water systems is 6 K (6 °C/42 °F supply, 12 °C/53 °F return temperature). At this temperature difference, the storage density of a chilled

Figure 5.12 The largest chilled water tank in the southern hemisphere at James Cook University in Queensland, Australia, has a capacity of 12.5 million litres. It is 17 m high and 30 m in diameter.

Source: James Cook University

water tank is approx. 7 kWh_{th}/m^3. Increased temperature differences increase the storage density linearly and reduce the water volume necessary to store a given cold load.

Chilled storage tank sizing – worked example

A building owner has the choice of installing either a fan-coil or a chilled-ceiling air-conditioning system. The fan coils require a supply/return temperature of 6/12 °C (42/53 °F), the chilled ceilings operate on a supply/return difference of 15/18 °C (59/64 °F). The building has a constant cooling load of 100 kW_r. A chilled water storage tank shall be installed to provide eight hours of chilled water. What are the tank volumes required for the two options?

Answer:
Eight hours of cooling at 100 kW_r of load equals 800 kWh_{th} of heat to be stored. The chilled water tank volume is calculated as per Equation (5.1). Therefore, for the fan-coil option, the tank volume is approx. 115 m^3. For the chilled ceiling option the volume is 230 m^3, or *twice as large* due to the reduced temperature difference (3 K instead of 6 K). Note that this example excludes the thermal loss of the tank and associated equipment. So, in reality, both tanks would have to be larger to account for this.

Chilled water tanks have a lower storage density than hot water tanks due to the lower temperature difference between supply and return to/from the tank. As mentioned above, they also have lower thermal losses than hot water tanks, as the temperature difference to ambient is usually smaller for these tanks. Chilled water tanks require insulation with a vapour barrier to prevent condensation on the cold tank surface and tank piping. Common tank materials are steel, concrete and plastic.

5.2.2 Ice and slurry

Ice storage tanks are frequently used and are state-of-the-art in conventional refrigeration systems. They make use of the latent heat of fusion of water (334 kJ/kg or 144 Btu/lb), which is greater than the specific heat capacity (4.2 kJ/(kgK) or 1 Btu/(lbF)) of liquid water at 0 °C (32 °F). This means that ice storage tanks only require approx. 20 per cent of the volume of a comparable chilled water tank to store the same amount of energy. They require supply temperatures below 0 °C (32 °F) during the ice-generation process, which excludes most standard air-conditioning chillers from being used. Low-temperature chillers with different refrigerants or multi-stage processes[1] are generally used instead.

[1] A multi-stage process uses two or more cascaded chillers. Each chiller reduces the supply temperature a little more. This combination allows lower supply temperatures to be reached than would be the case using one-stage chillers.

Ice storage tanks are suitable for use in solar air-conditioning systems; however, they do not work with the most common thermal chillers – water/LiBr ABsorption and ADsorption chillers, which have a chilled water limit of approx. 5 °C (41 °F). Below that, water freezing in the evaporator may cause damage in those chillers. The only thermal chillers suitable for ice storage are ammonia/water chillers with evaporator temperatures down to –30 °C (–22 °F).

There are several different technologies commercially available for ice storage tanks:

- *Coils with internal melt.* The storage tank, filled with water, has submerged coils in it. Cold coolant flows through the coils, causing ice to form on the outside of the coils. When the water in the tank is frozen, the refrigerant is replaced by warm coolant from the load. The warm coolant melts the ice, is thus chilled and returns to the load (see Figure 5.13).

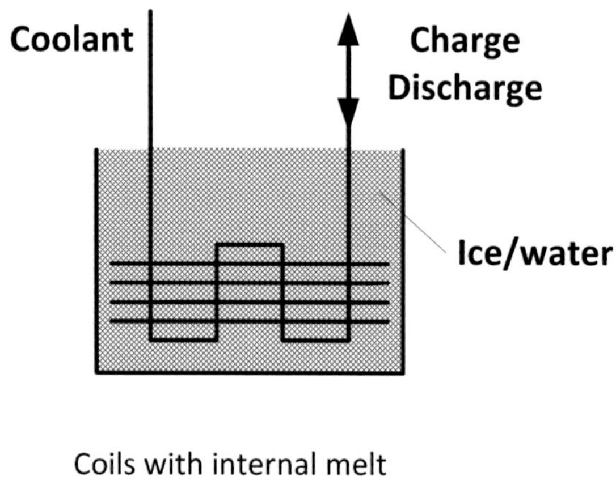

Figure 5.13 Coils with internal melt. Coolant is used for charging and discharging in a static water tank with ice formation on the outside of the coolant coils.

Source: Solem Consulting

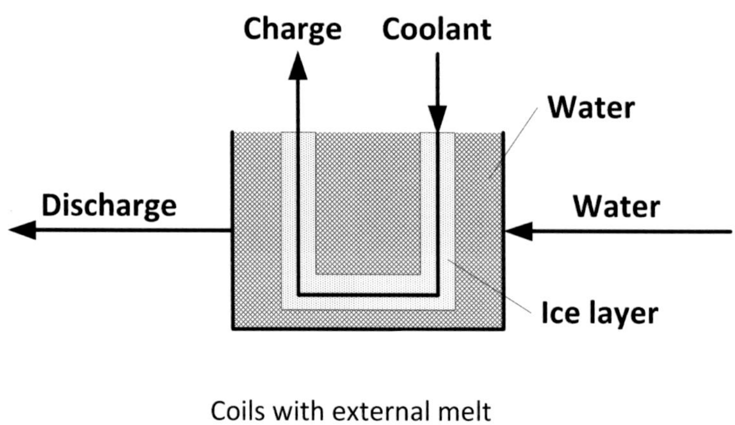

Figure 5.14 Coils with external melt. Coolant is used to charge the tank, i.e. to generate ice on the outside of submerged coils. Water is used for the discharge. It is pumped through the tank, thus melting the ice on the coils.

Source: Solem Consulting

- *Coils with external melt.* Submerged coils in a water tank are used with internal refrigerant or cold coolant flow, causing ice to form on the external surfaces of the coils. Before the space between the coils is completely frozen up, warm water from the load is pumped over the outside of the coils, melting the ice. The cooled water then returns to the load. Low-pressure air bubbles underneath the coil are sometimes used to agitate the water and assist the melting (see Figure 5.14).

- *Ice harvesting*. The evaporator of a refrigeration system is placed over a storage tank. It commonly consists of stainless steel plates welded into channels with internal flow of refrigerant. Externally, the plates are wetted with water and ice forms on both sides. At predetermined intervals the refrigeration process is reversed, the evaporator becomes the condenser and hot refrigerant enters the plates. This causes the ice to break away and drop into the tank (see Figure 5.15).

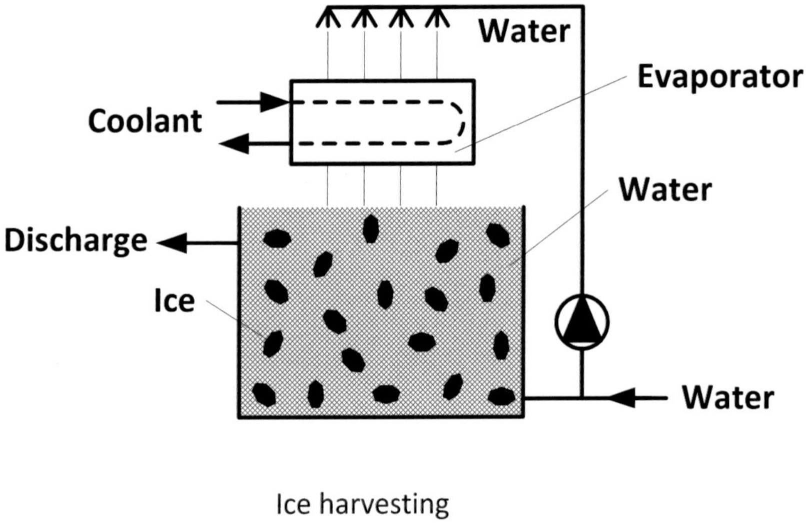

Figure 5.15 Ice harvesting. The evaporator of a refrigeration system is placed over a storage tank. Externally, the plates are wetted with water and ice forms on both sides. At predetermined intervals the evaporator is heated. This causes the ice to break away and drop into the tank.

Source: Solem Consulting

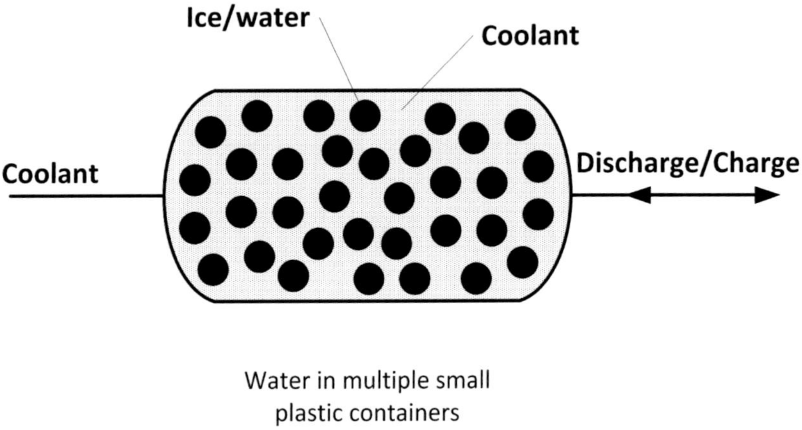

Figure 5.16 Multiple encapsulated plastic containers. Coolant is used for charging and discharging in a static tank filled with coolant and multiple encapsulated plastic containers.

Source: Solem Consulting

- *Water in multiple small plastic containers*. Water is encapsulated in small, spherical containers. The containers are submerged in a tank that is filled with coolant. When cold coolant is circulated through the tank, the water freezes inside the containers. Circulation of warm coolant melts the internal ice again (see Figure 5.16).

- *Slurry.* Slurry ice is a mixture of liquid water or water–glycol and micro ice crystals (typically 0.1–1 mm in diameter). A freezing point depressant is added to achieve temperatures below 0 °C (32 °F). Slurry ice is generated by constant wiping or scraping of a cold surface wetted with water. It can be pumped like water and stored in a tank. A cooling load can be met directly by pumping the slurry through the load or via a secondary heat exchanger (see Figure 5.17).

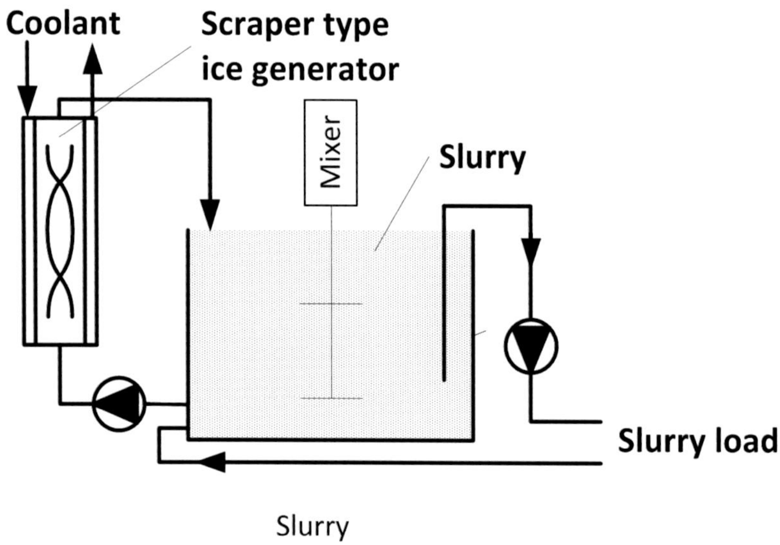

Figure 5.17 Slurry ice circuit. Coolant is pumped through an ice generator and ice forms on the cold walls. Slurry ice is generated using constant wiping or scraping of the cold surfaces wetted with water.

Source: Solem Consulting

5.2.3 Eutectic salts

The main advantage of ice storage is the compact storage size compared to chilled water storage. A disadvantage, however, is that standard air-conditioning chillers struggle to achieve the negative temperatures required to freeze water. Lower evaporator temperatures result in lower COPs and higher power consumption by the compressor. A solution is to raise the freezing point of water so that standard air-conditioning chillers with a supply temperature of 6 °C (42 °F) can freeze this water. The freezing point increase is achieved by mixing hydrated salts with water. Typically, the freezing point can be raised to approx. 8 °C (46 °F). This water–salt mixture is called eutectic salts storage or warm ice.

- *Eutectic salts.* The water–salt hydrate mixture is encapsulated in small plastic containers, submerged in a storage tank filled with water (Figure 5.18). The water is circulated through the tank, either to freeze or to melt the mixture. The phase change of freezing/melting results in a storage density comparable to pure ice, thus eutectic salt storage tanks can be rather compact. Standard air-conditioning chillers can be used to freeze the mixture. The supply temperature to the load is around 8–10 °C (46–50 °F), which limits this kind of tank to applications without supply air dehumidification.

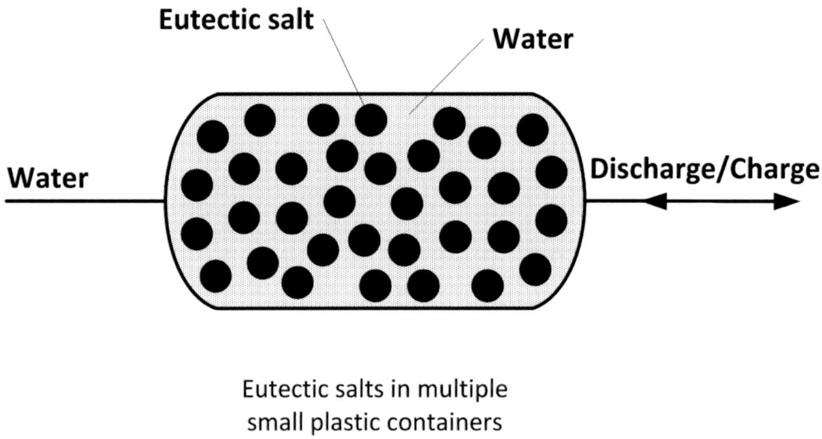

Figure 5.18 Eutectic salt cold storage. The water–salt hydrate mixture is encapsulated in small plastic containers, submerged in a storage.

Source: Solem Consulting

6

Designing and sizing solar cooling systems

This chapter deals with the designing and sizing of solar cooling systems. Different design approaches and methods are presented, as well as worked examples for three different climate locations. Different simulation software tools are presented in addition to the methods. Solar thermal cooling systems are discussed in greater detail in this chapter since complexity is greater than in PV-driven systems. General information on PV cooling can be also found in Chapter 3.5.

Three design approaches are presented in this chapter:

1. Rule-of-thumb method for a preliminary system design
2. Check-list method for a more advanced system design
3. Manual method for a detailed system design

Before the start

Before starting a solar cooling design the following questions need to be asked:

1. What is the orientation (e.g. south, south-west etc.) and size (m^2, sq.ft) of the building area available for mounting solar collectors or PV modules? Is there extra space available on the roof, the ground or on the façade for the installation of the above?
2. Does the building already have an existing active cooling system? If so, what is the capacity and type?
3. Are the cooling, heating and domestic hot water (DHW) requirements of the building known? If not, these must be evaluated.
4. How much space is available in the technical room and/or besides the building for the installation of a solar cooling system and storage tanks?

When starting with the design and sizing of a solar cooling system some general options have to be considered. In general, a building's cooling requirement is determined by solar radiation (heat through the building glazing), the emission of heat and humidity from people and heat-emitting equipment in the building (e.g. computers, machinery, etc.). The cooling requirement is the quantity of this heat, which for comfort reasons must be removed from the building. Active air-conditioning systems, which need electricity or heat to generate some form of cold, are typically used for this. But there are also other options to reduce the cooling load before designing and installing an active system. Heat from outside that can be initially prevented from entering the building does not have to be actively removed from the building via various measures. This is called passive cooling (see Chapter 1.1).

Free cooling should also be investigated as a further option, because it is an economical method of air-conditioning. Free cooling uses an installed heat rejection system (e.g. a wet cooling tower or dry cooler) at low ambient air temperature (see Figure 6.1) to provide chilled water for a building without using other active air-conditioning components. On days with low ambient temperature the cooling tower itself is sufficient to provide the required chilled water temperature for air-conditioning. Additional electric chiller operation is not required. This method saves electricity, since only the cooling tower and water pumps have to be operated. To use free cooling, an additional heat exchanger is required between the cooling water and chilled water circuit to separate the two. Without it, solids from the open circuit of the wet cooling tower could enter the chilled water distribution of the building. In the case of free cooling, the conventional (or thermal) chiller is not used. Energy savings of up to 75 per cent can be achieved against a conventional air-conditioning system [11].

In general, the following measures to reduce the cooling requirement of an existing or new building should be investigated and considered before starting with the design and sizing of a solar cooling system:

- shading of east, south and west-facing glass surfaces (northern hemisphere); or east, north and west-facing glass surfaces (southern hemisphere);
- use of heavy building materials/mass (e.g. concrete walls or ceilings);
- increased chilled water temperature to achieve the required comfort level, e.g. to 10 °C (50 °F) or 15 °C (59 °F) instead of 6–7 °C (42.8–44.6 °F) chilled water temperature.

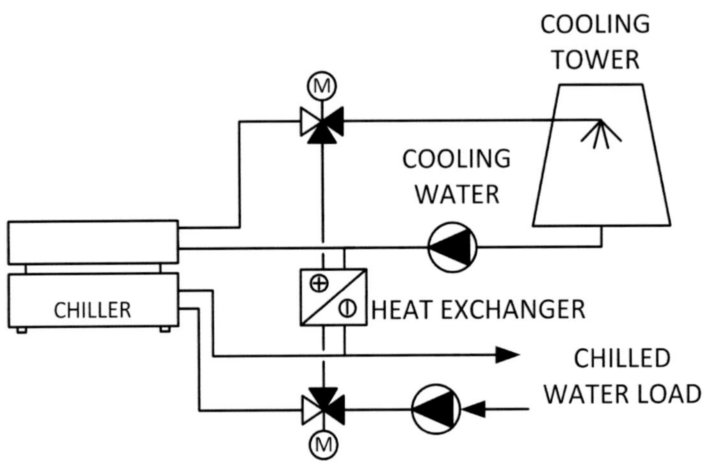

Figure 6.1 Example schematic of a free cooling system. Note that the cooling tower shown is exemplary. Free cooling works also with all kinds of heat rejection technologies (see Chapter 3.3.3). In general, free cooling is possible when ambient temperatures are sufficiently low.

Source: Solem Consulting

There are many possible configurations possible when it comes to system design, as shown in the following.

Overview of solar cooling system configurations for solar thermal or PV-driven, systems providing cooling only and systems providing both cooling and heating

Figures 6.2 and 6.3 show the different decision paths for recommended system configurations and their corresponding system components with respect to possible storage requirements and type, cold distribution system, solar technology and climate zone. Figure 6.2 shows the situation for a building that requires cooling only. Please note that recommendations given here are suggestions only. Other combinations are possible.

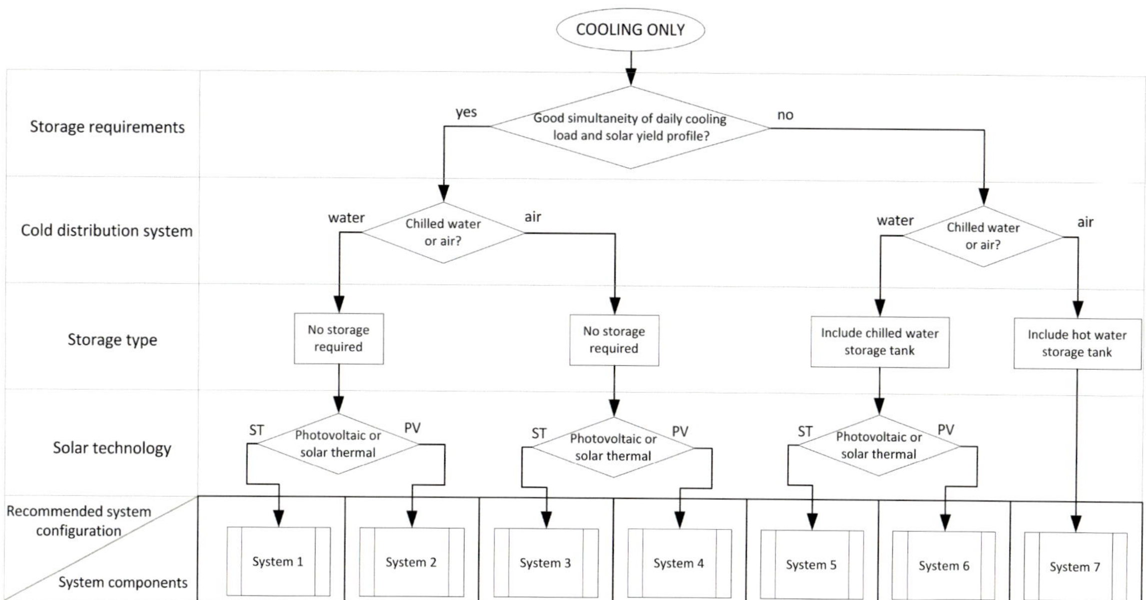

System components	System 1	System 2	System 3	System 4	System 5	System 6	System 7
Solar thermal collectors	X		X		X		X
PV panels		X		X		X	
AB/ADsorption chiller	X				X		
DEC			X				X
Hot water storage tank							X
Chilled water storage tank					X	X	
Gas boiler	X		X		X		X
Conventional chiller		X		X*		X	

(* plus additional air-water heat exchange in air handling unit)

Figure 6.2 Overview of recommended solar cooling system configurations for solar thermal or PV-driven; for cooling only.

Note: ST = solar thermal collectors, PV = PV modules

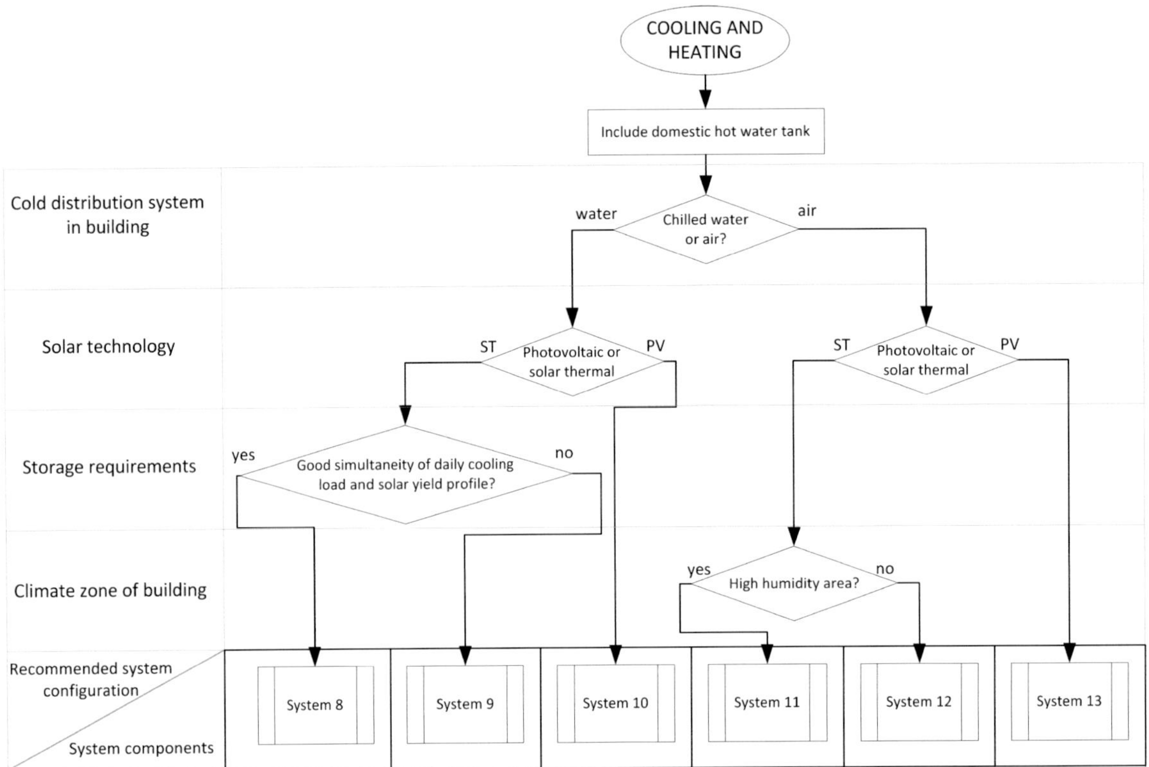

System components	System 8	System 9	System 10	System 11	System 12	System 13
Solar thermal collectors	X	X		X	X	
PV panels			X			X
AB/ADsorption chiller	X	X				
DEC				X	X	
Hot water storage tank	X	X	X	X	X	X
Chilled water storage tank		X				X
Gas boiler	X	X		X	X	
Conventional Air-Conditioner/Chiller			X	X		X*
Electrical heat pump			X			X*

(* plus additional air-water heat exchange in air handling unit)

Figure 6.3 Overview of recommended solar cooling system configurations for solar thermal or PV-driven; for systems providing cooling and heating.

Note: ST = solar thermal collectors, PV = PV modules

6.1 Rule of thumb method

The following rule of thumb method can be used to carry out a preliminary system design and sizing for solar thermal cooling systems. Starting with a given cooling requirement of a building (in W/m^2 of building floor area) and a given chiller capacity (in kW_r), an estimate of the required solar collector area size (in m^2) can be determined, as well as the corresponding hot water storage tank volume (in litres). The cost for a system determined in such a way can be calculated using specific cooling capacity costs (EUR/kW_r, see Chapter 8.2.2), yielding a total system investment cost. With this information one can assess whether a proposed solar cooling system has an economic benefit or not. The rule of thumb method is quick and easy to use before commissioning a detailed system design study. However, it has to be noted that the specific figures in Table 6.1 have an inaccuracy which is due to their empirical character. Component sizes derived with the rule of thumb method are only a preliminary assessment and should not be used to make executive design decisions. The rule of thumb calculations given in Table 6.1 are based on the analysis of worldwide installed solar cooling systems.

Worked example: solar cooling for a modern office building using the rule of thumb method for a preliminary system design

A modern office building with $2,000\,m^2$ (21,528 sq.ft) of office space needs to be cooled. It is assumed that the specific cooling load is $35\,W/m^2$ (11.1 Btu/hr/sq.ft) of building floor area. This number can be determined from the local building standard for office spaces. This gives a peak cooling load of $35\,W/m^2 \cdot 2,000\,m^2 = 70\,kW_r$, or 21,528 sq.ft · 11.1 Btu/hr/sq.ft = 238,850 Btu/hr.

Using the rule of thumb method given in Table 6.1, the specific collector for a medium-/large-scale system is chosen as $3.0\,m^2/kW_r$ (0.00946 sq.ft/hr/Btu). The size of the solar collector area is then calculated as $3.0\,m^2/kW_r \cdot 70\,kW_r = 210\,m^2$, or 0.00946 sq.ft/(hr Btu) · 238,850 Btu/hr = 2,260 sq.ft. The hot water storage tank is calculated to $210\,m^2 \cdot 50\,l/m^2 = 10.5\,m^3$, or 2,260 sq.ft · 1.23 gal/sq.ft = 2,772 gal. With this information the planner can decide whether there is enough roof space available for the solar collector area installation or not, and whether there is sufficient space for the hot water storage volume required.

Table 6.1 Rule of thumb calculations for preliminary solar thermal cooling system design and sizing

	AB-/ADsorption for medium-/large-scale systems	AB-/ADsorption for small-scale standardised systems	Open systems (DEC and liquid sorption)
Specific collector area per cooling capacity	3.0–3.5 m^2/kW (0.00946–0.01104 sq.ft/(hr btu))	4.5 m^2/kW (0.01419 sq.ft/(hr btu))	1.0–1.5 m^2/kW (0.00315–0.00473 sq.ft/(h btu))
Specific collector area per installed air volume flow	N.A.	N.A.	8–10 m^2 per 1,000 m^3/h (86–108 sq.ft per 35,315 ft^3/hr)
Hot water storage per installed collector area	50 l/m^2 (1.23 gal/sq.ft)	50 l/m^2 (1.23 gal/sq.ft)	50 l/m^2 (1.23 gal/sq.ft)

6.2 Checklist method

There are different approaches to designing a solar cooling system, but the most important decision is the appropriate choice of cooling technology (e.g. solar cooling system with ABsorption chiller or DEC system). We present a checklist method developed by the IEA-SHC Task 38 'Solar Air-Conditioning and Refrigeration' [12], which is available for download at www.tecsol.fr/checklist.

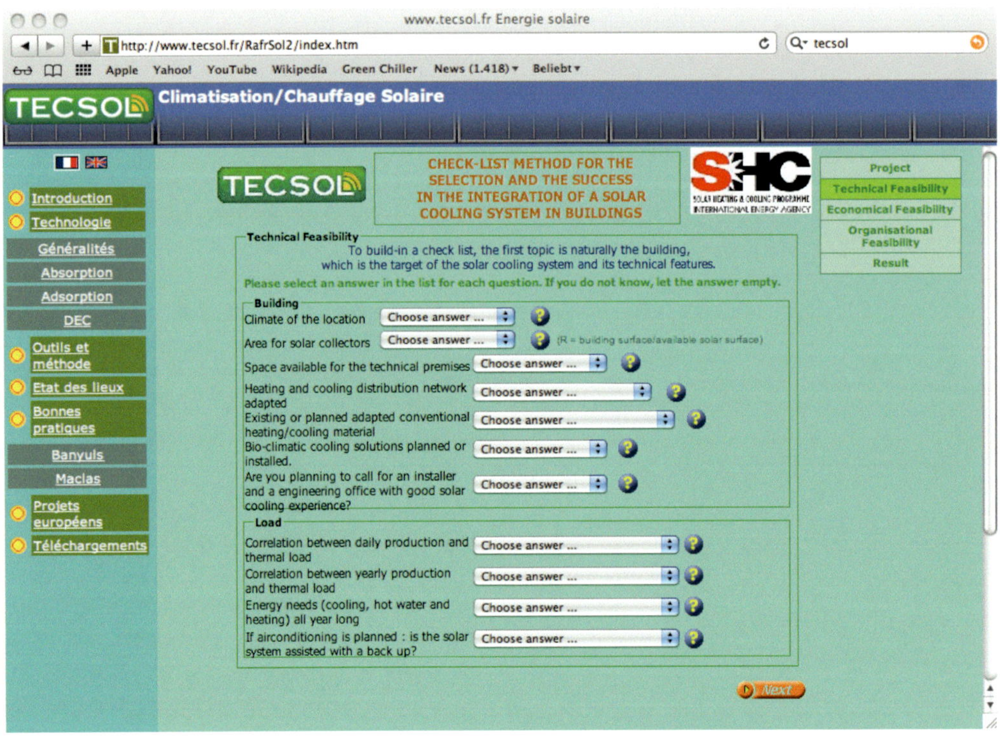

Figure 6.4 Screenshot of the IEA-SHC Task 38 advanced scheme checklist method. The method is web-based and needs to be completed online. Available at www.tecsol.fr/checklist.

Source: Tecsol

The checklist starts with a short project description (see Figure 6.4), followed by questions on the technical feasibility (e.g. building and loads), the economical feasibility (e.g. energy cost, information about the building owner) and the organisational feasibility (e.g. operation and maintenance, monitoring). The results of the analysis are then evaluated according to a 20-point scoring scheme. This preliminary analysis of the proposed system enables one to see whether the project is technically and economically feasible or not. If it is feasible, then the method can show the requirements for some of the project steps like components, planner, building owner, installer and monitoring of the installation.

The checklist method delivers a technical recommendation as well as an economic assessment. The feasibility of a planned project is rated and the requirements of important parameters are given.

6.3 Manual design method

This method involves a much higher level of calculation, but is also the most detailed of the methods presented in this chapter. It is explained using a general schematic of a solar thermal cooling system for ABsorption and ADsorption chillers (closed systems), as shown in Figure 6.5. The overall system consists of six individual sections which are hydraulically coupled. The chapters in this book corresponding to each system section are indicated.

Example system description

The solar collector section in Figure 6.5 includes the solar collector field with solar collectors, a primary solar collector pump P1 and a heat exchanger as the connection to the heat storage system section. The heat storage section contains a hot water storage tank with a secondary pump P2 and, in parallel, a conventional boiler as hot back-up heat supply with a hot water pump P3 to the thermal cooling section. The thermal cooling section includes the ABsorption or ADsorption chiller. In the heat rejection section the waste heat from the thermal chiller has to be rejected. This is done using a heat rejection unit (e.g.

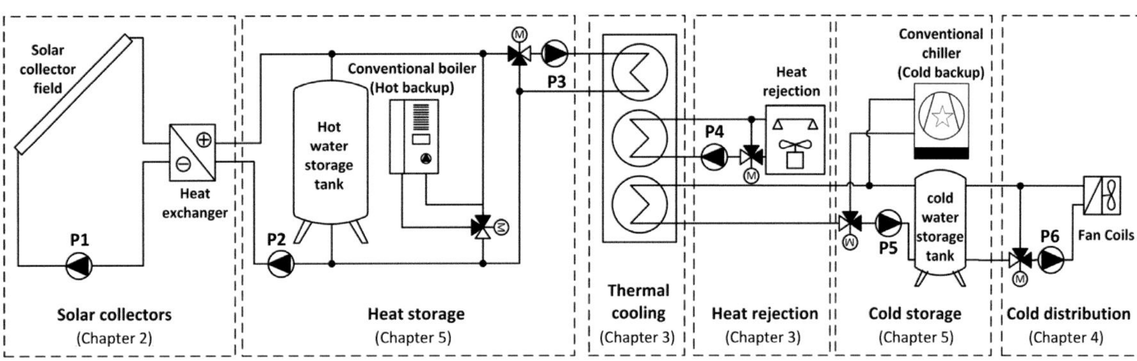

Figure 6.5 General solar cooling system schematic. Both options for hot and cold back-up, as well as storage, are shown. It is possible, but usually not required, to operate both options simultaneously in the same system.

Source: Solem Consulting

wet cooling tower or dry cooler) operating in ambient conditions; it is linked to the chiller by a cooling water pump P4. The cold storage section consists of a primary chilled water pump P5, a cold storage tank and, additionally, a conventional chiller as cold back-up. The last section represents the cold distribution, which includes a secondary chilled water pump P6 and the cold distribution in the building (e.g. fan coils, cooling ceilings).

Manual design methodology

A manual design methodology is shown in Figure 6.6. The complete design process is divided into seven sections, symbolised by boxes (red dashed lines), which refer to the building, climate data, chiller, heat rejection, solar collector and storage, control and a performance check approach, respectively.

Note: Figure 6.6 is a condensed overview of a rather complex process. Thermal losses (e.g. in the piping) or efficiency variations (e.g. of the solar collectors or heat rejection units) are not shown; but these need to be considered in final designs. This manual method can be used as a first design, e.g. in pre-feasibility studies. It can also be used for a preliminary cost assessment. Note: the manual method can also be applied to an open DEC system, in which case the chiller, cooling tower and fluid piping have to be exchanged with the DEC unit, heat exchangers and air ducts, respectively.

The methodology follows the following steps (see Table 6.2):

1. Determine the specifications of the annual cooling, heating and hot water load (each needs to be specified separately by a standard norm calculation or determined by a simulation software tool), respectively. This refers to the parameters of the group 'Building' in Table 6.2.
2. Determine the weather and radiation parameters from databases or local measurements. This refers to the parameters of the section 'Climate Data' in Table 6.2.
3. Determine nominal cooling capacity of the building *together* with the annual thermal COP of the chiller. This leads to the required thermal power for the peak cooling capacity of the chiller. This refers to the parameters of the group 'Chiller' in Table 6.2.
4. Determine further inputs like average annual collector efficiency, annual average daytime global irradiation and annual average global insolation on collector surface. These are required to calculate the total collector area and the hot water storage tank volume, respectively. This refers to the parameters of the section 'Solar Collectors and Heat Storage' in Table 6.2.
5. Implement a control strategy (see Chapter 6.3.6).
6. Determine the yields for the annual thermal energy supplied by the solar field to the chiller, the annual cooling energy provided by the chiller and the required thermal energy for the chiller. Further, heating and hot water load can be calculated, respectively. Based on these results the solar fraction for cooling or heating and hot water can be calculated to evaluate the amount of available solar energy for the different applications. Additionally, the total required back-up energy needs to be calculated. Determine the total auxiliary power consumption of the solar cooling system based on the individual components as listed. This refers to the parameters of section 'Performance Check' in Table 6.2.

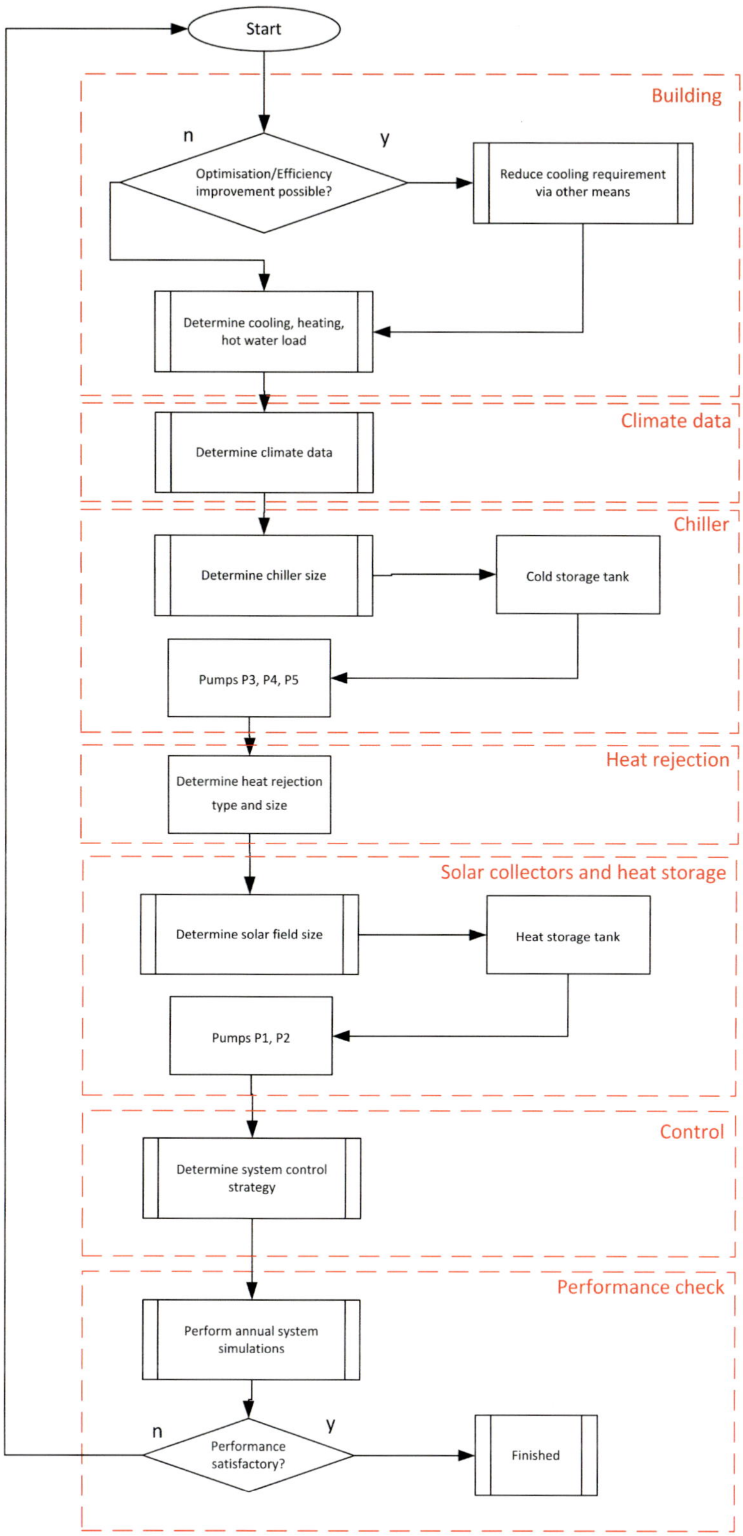

Figure 6.6 Manual design methodology for solar cooling systems. Sections are outlined in red.

Source: Solem Consulting

It is recommended the reader replicates Table 6.2 in a spreadsheet program and implements the calculation procedure illustrated.

In the following sections, the parts of Figure 6.6 are explained in more detail, including the manual calculations to determine required design parameters.

Table 6.2 Parameter calculation for manual design methodology; sections refer to Figure 6.6.

Section	Symbol	Calculation method
Building	$Q_{build,c}$	Annual cooling load of building – needs to be specified
	$Q_{build,h}$	Annual heating load of building – needs to be specified
	$Q_{build,hw}$	Annual hot water load of building – needs to be specified
Climate data	$G_{DN,titled,ann}$	Annual DNI radiation on collector surface
	G_{Design}	Design radiation for manual method
Chiller	$\dot{Q}_c$	This is the installed nominal cooling capacity of the chiller
	$COP_{th,ann}$	The annual thermal COP is the total cold energy per year divided by the total heat input per year. It can be estimated from the nominal COP given by the manufacturer. A reduction of 10–30 per cent of the nominal COP should be assumed, accounting for start-up/shutdown and thermal losses of the chiller and its piping.
	$\dot{Q}_{th}$	$\dot{Q}_{th} = \frac{\dot{Q}_c}{COP_{th,ann}}$ Thermal power required for nominal cooling capacity of chiller
Heat rejection	$\dot{Q}_{CW}$	$\dot{Q}_{CW} = \dot{Q}_C + \dot{Q}_{th}$ Thermal power of heat rejection unit (wet/dry cooler) required for nominal operation of chiller
Solar collectors and heat storage	$\eta_{coll,ann}$	The average annual collector efficiency is the total heat collected per year divided by the total solar irradiation on the collector. It can be estimated from the collector efficiency at the operating temperature of the chiller. A reduction of 5–15 per cent of that is recommended to account for thermal losses of the collector circuit
	$E_{g,surf}$	Annual average daytime global irradiation on collector surface can be determined from weather data at project location
	$G_{g,surf}$	Annual average global insolation on collector surface can be determined from weather data at project location. Note that this number represents solar input on the aperture surface of the collector, not on the horizontal
	$A_{coll,req}$	$A_{coll,req} = \frac{\dot{Q}_{th}}{\eta_{coll,ann} \cdot E_{g,surf}}$ Collector area required
	f_{os}	The collector field can be oversized to meet solar fraction criteria, otherwise set this value to zero
	$A_{coll,tot}$	$A_{coll,tot} = A_{coll,req} \cdot (1 + f_{os})$ Total collector area
	$V_{storage}$	$V_{storage} = \frac{\dot{Q}_{th} \cdot t}{\rho_{hw} \cdot c_{p,hw} \cdot \Delta T_{storage}}$ Hot water storage tank volume for peak cooling capacity

Section	Symbol	Calculation method
Control		Implement control strategy as per Chapter 6.3.6.
Performance check	Q_{th}	$Q_{th} = G_{g,surf} \cdot \eta_{coll,ann} \cdot A_{coll,tot}$ Annual thermal energy supplied to chiller
	Q_c	$Q_c = Q_{th} \cdot COP_{th,ann}$ Annual cooling energy provided by chiller as product of annual thermal energy supplied to chiller multiplied with annual thermal COP
	n_{op}	$n_{op} = \frac{Q_c}{\dot{Q}_c}$ Full load operation hours per year as ratio of annual cooling energy to nominal cooling capacity
	$Q_{th,req,c}$	$Q_{th,req,c} = \frac{Q_{build,c}}{COP_{th,ann}}$ Thermal energy required for cooling load of building as ratio of annual cooling load of building to average annual collector efficiency
	$Q_{th,req,hhw}$	$Q_{th,req,hhw} = Q_{build,h} + Q_{build,hw}$ Thermal energy required for heating and hot water load of building
	SF_c	$SF_c = \frac{Q_c}{Q_{build,c}}$ Solar fraction of cooling
	SF_{hhw}	$SF_{hhw} = \frac{Q_{th} - Q_{th,req,c}}{Q_{th,req,hhw}}$ Solar fraction of heating and hot water
	Q_{backup}	$Q_{backup} = Q_{th} - Q_{th,req,c} - Q_{th,req,hhw}$ Total back-up energy required
	η_{backup}	Back-up heater efficiency – needs to be specified
	$Q_{backup,PE}$	$Q_{backup,PE} = \frac{Q_{backup}}{\eta_{backup}}$ This is the primary energy demand for the back-up heater, e.g. the calorific value of the amount of gas required. It can be used to calculate the back-up energy cost
	P_{solar}	Power consumption of solar circuit pump – needs to be specified as per design
	P_{hw}	Power consumption of hot water circuit pump – needs to be specified as per design
	P_c	Power consumption of thermal chiller – needs to be specified as per design
	P_{cw}	Power consumption of cooling water circuit pump – needs to be specified as per design
	P_f	Power consumption of heat rejection unit fan – needs to be specified as per design
	P_{chw}	Power consumption of chilled water circuit pump – needs to be specified as per design
	P_{ctrl}	Power consumption of control and monitoring equipment – needs to be specified as per design
	$P_{aux,tot}$	$P_{aux,tot} = P_{solar} + P_{hw} + P_c + P_{cw} + P_f + P_{chw} + P_{ctrl}$ Total auxiliary power consumption as sum of all power consumptions
	$COP_{el,sys}$	$COP_{el,sys} = \frac{\dot{Q}_c}{P_{aux,tot}}$ Electrical COP of total system at nominal operation

6.3.1 The building

The first step in any cooling system design process should always be to attempt a reduction of cooling requirement by the use of passive methods. Furthermore, the efficiency improvement possibilities for the building should be checked and the cooling requirements minimised as much as possible. There is a big difference if the system is being designed for an existing or new building. In a new building, for example, the air-conditioning system to be installed can be optimised – more efficient cooling ceilings instead of air-handling units can be chosen – thus improving the overall performance of the system. In existing buildings, the cold distribution and the chilled water temperatures are very often fixed, adding restrictions in design choices. In all cases, the climate (outside ambient temperature, wet bulb temperature, relative humidity), the building standard (level of insulation, heat and cold distribution) and the cooling, heating and hot water loads need to be determined or estimated.

6.3.2 Climate data

A weather and radiation analysis needs to be carried out to determine the local annual maximum air temperature (dry ambient air temperature), the local

Climate: selection rules for heat rejection technology

The selection of the appropriate heat rejection technology mainly depends on the climate conditions at a project location. General rules are as follows:

- *Wet cooling towers*: The minimum achievable cooling water supply temperature is approx. 3 K higher than the given *wet* bulb temperature of ambient air. Wet cooling towers provide better system performance if ambient air *humidity* is low.
- *Dry coolers*: The minimum achievable cooling water supply temperature is approx. 3–5 K higher than the given *dry* bulb temperature of ambient air, depending on the quality of the cooler. They provide better system performance if ambient air *temperature* is low.

Generally, wet cooling towers can provide lower cooling water supply temperatures than dry coolers (assuming ambient humidity is low). Hot and humid regions are less suitable for wet cooling towers, as the evaporation of cooling water into ambient air is limited. There, dry coolers are the preferred option.

Note: The above rules are based solely on thermodynamic aspects. Investment and operational cost of the above heat rejection technologies are important for the economics of a solar cooling system. It is recommended to perform annual simulations once a heat rejection technology has been chosen. Simulations will provide more detailed results on which heat rejection technology is to be chosen.

annual mean maximum and mean minimum temperature and the wet bulb temperature (necessary for wet cooling tower design). Furthermore, the design radiation level for the location has to be determined (see Chapter 6.3.8). Solar radiation and weather data with hourly resolution is available from weather data software such as Meteonorm or EnergyPlus (see Chapter 12.6 for more databases).

6.3.3 Chiller

Once the cooling and heating loads of the building are determined, the next step is to determine the size of the AB-/ADsorption chiller (nominal cooling capacity, COP) either for peak design (maximum required cooling capacity) or base load design (annual average cooling capacity). See the following box.

To size an AB-/ADsorption chiller the required operating conditions (chilled and cooling water temperature) need to be evaluated first. This determines the hot water (heat input) of the chiller (see Chapter 3.2). The next step is the selection of the correct chiller size. It should be chosen according to the cooling capacity it will be operating at most often throughout the year (usually not the peak cooling load of the building – see Box above). This cooling capacity can be determined by an annual energy yield analysis based on system simulations.

Chiller: design approaches

There are two fundamentally different approaches to designing a solar cooling system with regard to the chiller. The first approach is simply to design the system to cover the peak cooling demand of the building. This means that the chiller is sized to the maximum cooling requirement of the building. In consequence, this chiller will be oversized during periods when the cooling requirement of the building is lower than its peak capacity. As a consequence, the solar system will also be oversized, incur more cost and be more liable to go into stagnation (see Chapter 7.2.1). Furthermore, the heat losses from the piping (e.g. between solar collector, heat storage and sorption chiller), as well as from the heat storage, are usually not accounted for in this design approach.

The second approach is a load-matching approach using annual energy yield analysis based on software simulations (e.g. TRNSYS) to predict the system performance over the year (utilising monthly or hourly values). The chiller capacity is then determined based on the correlation between building cooling load profile and the number of hours each cooling load is supplied by the system. Cooling capacity is chosen to cover most of the base load of the building, but not the peak loads. This design approach is much more time-consuming and requires more expert knowledge. This is mainly because the major system components have to be physically modelled in the simulation program, as well as the control philosophy. The heat losses between the single components are accounted for in this approach.

Then, depending on the chosen collector type, the chiller can be chosen. Low-temperature collectors such as flat plate, evacuated tube or CPC collectors are generally used with single-effect AB-/ADsorption chillers; high-temperature collectors such as parabolic trough and linear Fresnel collectors can be used with a double- or triple-effect ABsorption chiller.

Once the chiller cooling capacity has been chosen, the heat rejection method and size (wet cooling tower, dry cooler, hybrid cooler, geothermal, etc.) needs to be decided. Based on that the pumps and piping for hot water distribution and for cooling and chilled water can be specified, respectively.

Chilled water and cooling water temperature

Typically, the chilled and cooling water temperatures of AB-/ADsorption chillers are set to meet standard air-conditioning criteria. There, chilled water temperatures are set to an inlet/outlet temperature of 12/6 °C (53.6/42.8 °F) in order to allow for dehumidification of supply air via sub-cooling below the dew point (the dew point is defined as the temperature at which an equilibrium state of condensing and evaporating of water on a surface is achieved at given humidity). In standard air-conditioning systems, cooling water is chosen to have inlet/outlet temperatures of 27/33 °C (80.6/91.4 °F) to provide nominal operation conditions of the chiller. However, increasing or decreasing the cooling water temperature will result in improved chiller operation conditions.

If the chilled water temperature and/or the cooling water temperature are adjusted to different design values instead of to the standard values, then the cooling capacity of the ABsorption or ADsorption chiller can be increased. A decrease of heat rejection temperature by 1 K equals a 6–7 per cent cooling capacity gain for ABsorption chillers (water/lithium bromide and ammonia/water). For ADsorption chillers (water/silica gel and water/zeolite) a decrease of heat rejection temperature by 1 K equals a 4–5 per cent cooling capacity gain. Furthermore, an increase of the chilled water temperature by 1 K equals a 4–5 per cent cooling capacity gain for ABsorption chillers (water/lithium bromide and ammonia/water), as shown in Figure 6.7. This is also similar for ADsorption chillers.

Chilled water distribution

The chilled water distribution, with or without a chilled-water storage tank, is usually the same as for conventional chillers (see Chapter 4). The temperature difference of inlet and outlet (ΔT) in the chilled water circuit is usually between 3 K and 6 K. A higher return temperature works better for the sorption chiller, which leads to an increased COP and available cooling capacity, as shown in Figure 6.7.

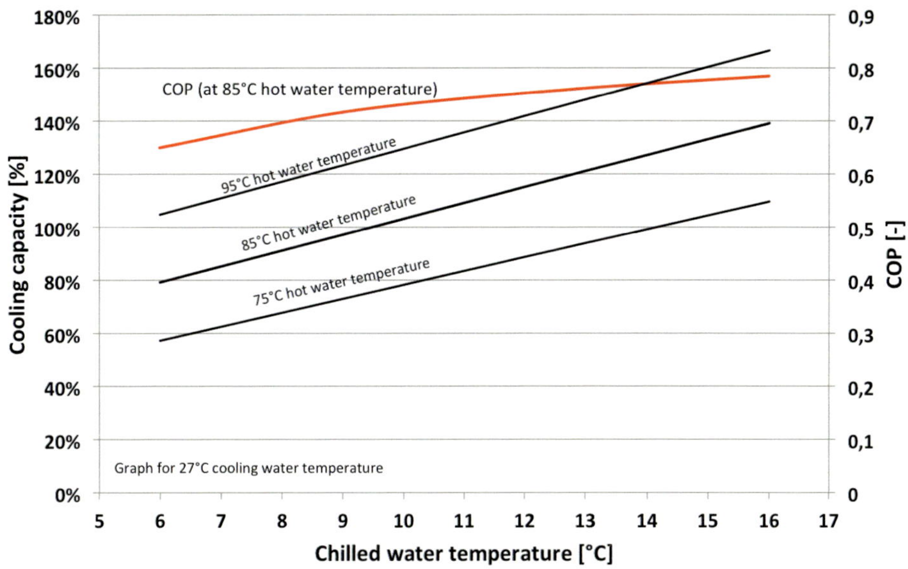

Figure 6.7 Example data for a 10 kW_r single-effect LiBr/water ABsorption chiller showing the cooling capacity and COP against chilled water temperature. It can be seen that with increasing chilled water temperature the cooling capacity for constant hot water temperatures also increases. Alternatively, the hot water temperature can be reduced for the same chiller capacity.

Source: [13, 14]

6.3.4 Heat rejection

Water treatment

Wet cooling towers have continuous evaporation of water during operation. A floating valve connected to make up water in the sump is typically used to automatically replace this water loss. If mains water is used, salts and solids dissolved in mains water will be introduced into the heat rejection circuit. With continuous water evaporation these salts and solids concentrate in the remaining water and eventually result in a sediment layer deposited in the cooling tower sump, requiring regular cleaning and maintenance (blow-down). Manual blow-down can be avoided if automated water treatment is applied to the mains water before it enters the cooling tower. The installation of a water treatment plant, which adds chemicals to the mains water, is recommended. These chemicals have the following functions: to increase solubility of water for salts, to hinder algae and bacteria formation and to control the pH value and thus provide corrosion protection.

However, even with chemical treatment, the salts and solids still concentrate in the circulating water, but this can be remedied via a solenoid valve. The concentration of these solids can be determined by measuring the electrical conductivity of the circulating water. Upon reaching a concentration limit the solenoid valve is opened to drain a defined amount of water, thus reducing the salt mass in the water circuit. This is automatically initiated through the water treatment plant controller.

Frost protection

If a cooling tower is located in frost-prone areas then frost-protection will have to be implemented. One option is to use a water–glycol mixture in the cooling water circuit. If this is not desired, then another option can be applied to wet cooling towers. The cooling tower sump (lower part of the cooling tower where the remaining cooling water is collected before it flows back to the chiller) can be arranged above a catchment tank in a frost-proof room below the cooling tower. Hydraulic piping is laid in such a way that the sump automatically drains into the catchment tank if the cooling water pump is switched off (see Figure 6.8). Compared to the water–glycol option this is a more complex option with higher cost. The cooling water pump has to be selected with a greater pump head to account for the extra static height. Some systems use electrical heating of the cooling tower sump. This is not recommended for solar cooling systems because of the additional power consumption.

Another important aspect of water/lithium bromide ABsorption chillers is that manufacturers typically specify a critical cooling water temperature of 25 °C. If the cooling water temperature falls below 25 °C (e.g. at low ambient temperatures) then the lithium bromide inside the ABsorption chiller may crystallise. This can lead to internal damage to the chiller. Commercial chiller controllers typically have built-in detection of low cooling water temperature but will simply initiate chiller shutdown in this case. To avoid frequent shutdown due to low cooling water temperatures, a three-way mixing valve in the heat rejection circuit is recommended (see Figure 6.5).

Fan power

In standard air-conditioning systems the fan of the heat rejection device usually runs at constant speed and is either on or off. This results in rather high power consumption, even if the heat rejection load is rather low; for example, due to lower ambient/wet bulb temperatures (the wet bulb temperature is the lowest temperature that can be achieved by evaporative cooling at 100 per cent relative humidity). A solar cooling system should operate as energy-efficiently as possible; hence a variable speed drive (VSD) attached to the fan motor of the heat rejection unit is recommended. It is one of the most important devices for effective and

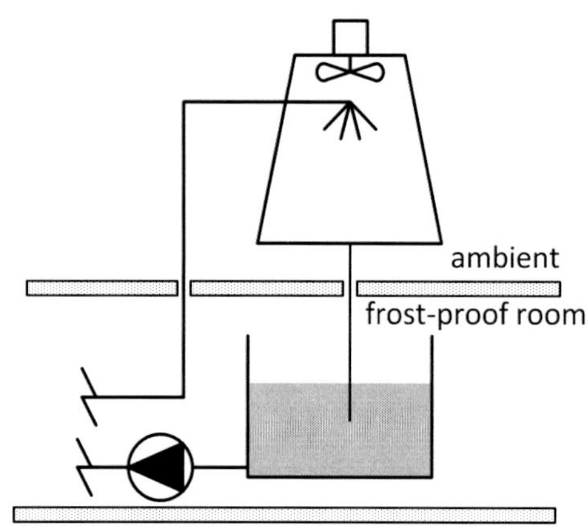

Figure 6.8 Cooling tower sump in a frost-safe room, an alternative method to the water–glycol option for frost protection of the wet cooling tower.

Source: Solem Consulting

energy-saving solar cooling system operation. It allows speed control of the fan for both wet cooling towers and dry coolers. The fan speed can be adjusted according to the load and ambient conditions. Fan power is proportional to the cube of the fan speed, thus a reduction in fan speed by 20 per cent results in a power reduction of nearly 50 per cent! Therefore, using a VSD, efficient fan operation can be implemented, e.g. for part-load operation of the chiller or during lower ambient temperatures. If free cooling is being used, the fan speed control provides an efficient means of temperature control of the supply water.

A small device saves cost

A VSD attached to the fan motor of the heat rejection unit is one of the most important devices for effective and energy-saving solar cooling system operation. Almost half of the fan power consumption can be saved if fan speed is reduced by only 20 per cent. Therefore, using a VSD, efficient fan operation can be implemented, e.g. for part-load operation of the chiller or during lower ambient temperatures.

6.3.5 Solar collectors and heat storage

The next section of Figure 6.5 to address is the solar collector section; its size can be determined by the required heat input for the AB-/ADsorption chiller. With this information the hot water storage tank size, the solar primary and secondary pumps (P1, P2) and the piping can be specified.

Volume flow

The hydraulic layout of a solar thermal collector field depends on the chosen AB-/ADsorption chiller and the required temperature difference between inlet and outlet (ΔT) of the hot water circuit. A low-flow (10–20 l/(h m^2)) volume flow rate is set for a high ΔT of 10–30 K. A high-flow (20–50 l/(h m^2)) volume flow rate is set if a low ΔT of 5–10 K is required. It is recommended that the solar collector pumps (P1, P2) are speed controlled (e.g. via a VSD) to maintain a constant hot water supply temperature during periods of changing irradiation. This is to achieve a stable operation of the chiller.

Tichelmann method: balanced flow in the solar field

The design of the collector-field piping can be done according to the Tichelmann method. This establishes a balanced flow in the solar field, i.e. the same volume flow rate through each solar collector (see Figure 6.9). In order to achieve this, the piping length of the return flow (cold side) needs to be twice as long as that of the supply flow (hot side) piping in order to maintain the same piping length for each collector string. It is highly recommended, especially for larger solar fields with more than ten collectors, to use balanced flow (Tichelmann) for the solar circuit in order to achieve the same flow rates through each string. If installed properly, balancing valves in each string are not needed. Albert Tichelmann (1861–1926) was a German engineer and expert in the area of hot water systems.

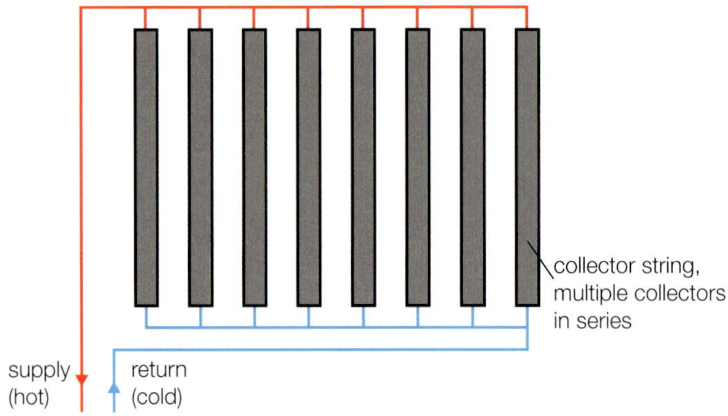

Figure 6.9 Tichelmann hydraulics.

Heat exchanger

A primary heat exchanger between the solar and heat storage circuit (see Figure 6.5) is not required if the heat transfer fluid through the solar collectors is the same as the fluid through the chiller. This is usually only the case in locations without frost. In frost-prone areas, the whole storage tank needs to be filled with a water–glycol mixture if no heat exchanger is used. This is usually more expensive than installing an additional heat exchanger. The heat exchanger provides a fluid divider between the solar circuit and the heat storage part, but involves temperature (~3–5 K) drop and pressure drop. If a heat exchanger is installed, an additional secondary pump (P2) is needed (see Figure 6.5).

Pumps

The correct choice of pumps is subject to standard engineering procedures and is beyond the scope of this book. However, it is highly recommended to use speed-controlled pumps for the solar circuit in order to maintain a constant supply temperature during changing irradiation conditions. Thermal chillers operate more efficiently with constant inlet temperatures due to their rather large thermal mass. Speed control also ensures energy-efficient pump operation. A VSD pump is recommended for the solar circuit and/or hot water storage circuit (pumps P1, P2 and P3 in Figure 6.5). Pumps with a built-in VSD can be purchased off-the-shelf.

Heat storage

The heat storage tank provides a heat buffer during low radiation periods. It is highly recommended to integrate the heat storage tank in a parallel configuration, as shown in Figure 6.5. In a parallel configuration, the direct operation of the chiller on solar heat is possible. In this case pumps P2 and P3 operate at equal flow rates and the heat storage tank is bypassed. This way there are no heat losses through the heat storage tank. If pumps P2 and P3 operate at different flow rates the tank is either charged with hot water from the solar collectors or discharged of hot water to the chiller. Parallel configuration is also recommended for other heat sources, such as the boiler back-up (see Figure 6.5), which allows separate charging/discharging/chiller operation. The placement of sensors at the heat storage tank is important for the control strategy (see Chapter 7.2.2).

6.3.6 Control

The control of a solar cooling system is a very important issue, because the system efficiency highly depends on that. If single controllers for each control loop are used, then the overall system efficiency is likely to be lower than if an overall master system controller is used. This is because multiple single control loops do not always operate together in the optimum combination. A master controller combines different control loops into a single controller unit. It is recommended for solar cooling system control.

Closed loop system control

Figure 6.10 describes the general simple control approaches for closed-loop systems incorporating ABsorption and/or ADsorption chillers. The system includes four control loops – C1, C2, C3 and C4 – each with a different function. These control loops are usually programmed in a PLC (programmable

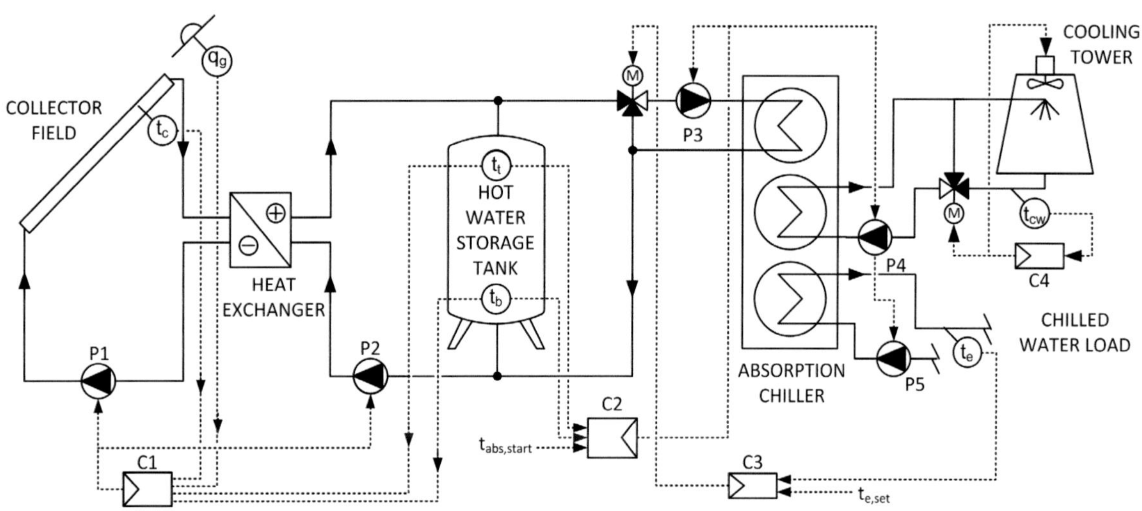

Figure 6.10 Control example of a complete closed-loop solar cooling system with four different control loops. These should be operated through a master controller unit.

Source: Solem Consulting

logic controller). PLCs are commercially available. Alternatively, a preconfigured system controller designed especially for solar cooling is offered by manufacturer SolarNext from Germany.

The control loops in Figure 6.10 are symbolised by the respective controllers C1 to C4. Controller C1 is responsible for the solar collector circuit, and controls the mass flow of pumps P1 and P2 to maintain a constant temperature difference between collector and storage tank. Therefore, the collector outlet temperature (t_C), as well as the hot water storage tank temperatures at the top and the bottom of the tank (t_t and t_b), are measured using temperature sensors. However, in some systems radiation measurements (q_g) for variable mass flow of the solar pump P1 with linear correlation to the incident radiation are used. The choice of control strategy (temperature or radiation-based) depends mainly on the solar collector field size. The larger the collector field the more thermal mass and slower reaction time. Radiation-based control is typically recommended for collector fields smaller than $100\,m^2$. Field sizes larger than that should use temperature-based controls.

The second controller, C2, controls the mass flow to the ABsorption or ADsorption chiller via pump P3 depending on the hot water storage tank temperatures (t_t and t_b). A start-up temperature is set for the chiller ($t_{abs,start}$). Pump P3 and the chiller are switched on if the hot water storage tank temperature at the top of the tank (t_t) has reached that start-up temperature. In that case, the same control command is also given to pump P4 (to start the cooling water flow) and to pump P5 to start the chilled water flow.

The chilled water control (controller C3) requires a temperature sensor in the chilled water circuit (t_e) as well as a set point for the required chilled water flow temperature ($t_{e,set}$). It is used to control the chilled water temperature via a mixing valve in the hot water circuit that controls the hot water temperature into the chiller. The correlation between hot water and chilled water temperature of an AB-/ADsorption chiller is given in Figure 6.7.

Controller C4 uses a temperature sensor to measure the cooling water temperature (t_{cw}) in the return flow of the heat rejection circuit. It has two tasks. First, C4 controls a three-way mixing valve in the heat rejection circuit to adjust the cooling water return temperature to the chiller. This is used to avoid low cooling water temperatures during chiller operation (see Chapter 6.3.3). Second, C4 controls the fan speed of the heat rejection unit to provide the required cooling water supply temperature.

Note: the above control strategy does not include any further logical operations that are necessary for the system operation. It is a guide for the key control loops of a solar cooling system. Logical operations such as system start-up or shutdown and alarm reactions are not included since these are common engineering practice. Also not included are instructions for data monitoring and logging.

The development and implementation of the control strategy for a solar cooling system is a complex task that requires expert knowledge, both in controls and measurement technologies. To avoid this, the authors recommend using a pre-configured controller for solar cooling systems, e.g. from German manufacturer SolarNext AG.

Open loop system control

Open systems such as DEC systems (see Figure 6.11) and liquid sorption systems (see Chapter 3.2.2) typically have four control sequences concerning the different operation modes of the system [15].

The control sequences are:

1. *Free ventilation mode*. None of the components except for the fans are active; no heat input is required. Fresh air can be used if ambient temperature and humidity are within comfort levels.
2. *Indirect evaporative cooling mode*. The return air evaporative cooler is active as well as the heat exchanger wheel. The return air is brought close to saturation and then enters the heat exchanger wheel. Thus, only sensible cooling of the supply air stream is provided. No heat input is required. Main control parameters: efficiency of return air evaporative cooler (0–100 per cent) through control of volume flow of sprayed water.
3. *Combined evaporative cooling mode*. The supply air and the return air evaporative coolers are active. The heat exchanger wheel is in operation. Combined evaporative cooling (direct and indirect) is used. No heat input is required. Main control parameters: efficiency of supply air evaporative cooler (0–100 per cent).
4. *Desiccant cooling mode*. The dehumidification wheel, the evaporative coolers, the heat exchanger wheel and the solar energy heat exchanger are active; all heat available from the solar system is used for regeneration of the desiccant wheel. Main control parameters: regeneration air temperature by means of control of the return fan rotational speed, supply air evaporative cooler (0–100 per cent).

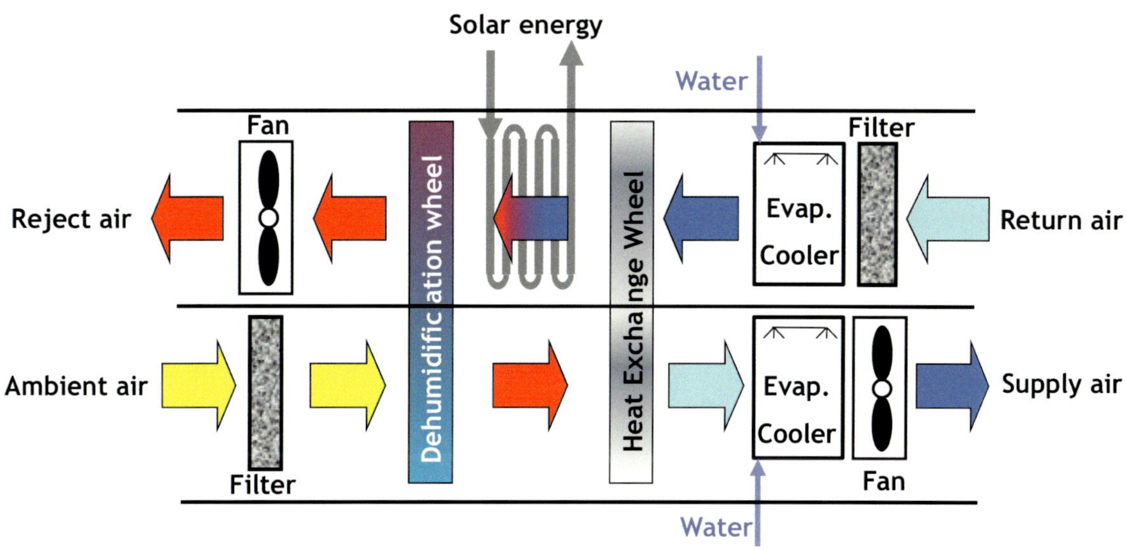

Figure 6.11 DEC system combined with solar energy as heat source for the regeneration of the dehumidification wheel.

Source: Solem Consulting

Such control sequences are usually programmed in a PLC controller, which is often connected to a building management system (BMS).

6.3.7 Performance check

Finally, a performance check is made, either using the checklist method described in Chapter 6.2 or using software simulation tools to model the system's annual performance. If the calculated performance is not satisfactory then the system design should be changed/modified. In order to do so, the manual design method can be repeated using different component sizes.

6.3.8 Worked example

The manual design method is now used for the design of three example systems to show different design solutions. This includes using different sorption technologies (closed and open systems) and solar collector types in three different types of buildings and in three different climates:

1. System No. 1 is a *residential building* in a *moderate climate* (location Munich, Germany) with a cooling capacity of 10 kW_r (2.8 RT). The system consists of closed-loop, single-effect ADsorption chiller, dry cooler and evacuated tube collectors with a heat storage tank.
2. System No. 2 is an *office building* in a *hot and dry climate* (location Abu Dhabi, UAE) with a cooling capacity of 34 kW_r (9.7 RT). A closed-loop, double-effect ABsorption chiller is used together with a wet cooling tower and small-size parabolic trough collector with a heat storage tank.
3. System No. 3 is a *school building* in a *hot and humid climate* (location Singapore) with an air volume flow of 10,000 m^3/h (35,315 ft^3/hr) using an open liquid DEC system together with flat plate collectors and a brine storage tank.

System No. 2 will be used as the main worked example in which each section of Figure 6.6 is explained in more detail. However, results are presented for all three example buildings.

Building

It is assumed the building has no further optimisation potential for the reduction of air-conditioning requirements.

Climate data

The Meteonorm database for Abu Dhabi has been used (www.meteonorm.com). Detailed weather data analysis for System No. 2 (Abu Dhabi) is given in Figure 6.12; it shows the dry ambient temperature and wet bulb temperature for that location over the year.

It can be seen in Figure 6.12 that the annual maximum ambient temperature and annual maximum wet bulb temperature for Abu Dhabi is approximately

45 °C (113 °F) and 29 °C (84.2 °F), respectively. The heat rejection technology can now be chosen as a function of these temperatures for the solar cooling system. Taking the selection rules given in Chapter 6.3.2, the options are as follows:

- Wet cooling tower: for the highest annual wet bulb temperature of 29 °C (84.2 °F) the cooling water supply temperature provided would be 32 °C (89.6 °F). This is because the cooling water supply temperature of a wet cooling tower will always be approximately 3 K higher than the wet bulb temperature. During the summer months the average wet bulb temperature in Abu Dhabi is typically at or above 25 °C (77 °F), leading to a cooling water supply temperature between 28 °C (82 °F) and 32 °C (89.6 °F).
- Dry cooler: for the highest annual wet bulb temperature of 45 °C (84.2 °F) the corresponding cooling water supply temperature provided would be approximately 48 °C (118.4 °F). Again, 3 K temperature difference is assumed between ambient dry air and cooling water supply temperature. For the summer months the average ambient air temperature in Abu Dhabi is typically at or above 35 °C (95 °F), leading to a cooling water supply temperature between 38 °C (100 °F) and 48 °C (118.4 °F).

As a consequence the wet cooling tower is chosen, because it provides lower cooling water supply temperatures over the year. This is favourable for the performance of the ABsorption chiller: a decrease of cooling water supply temperature by 1 K equals a 6–7 per cent cooling capacity gain for ABsorption chillers (water/lithium bromide and ammonia/water – see Chapter 6.3.3). Therefore, the annual cooling yield of the chiller will be increased by choosing a wet cooling tower.

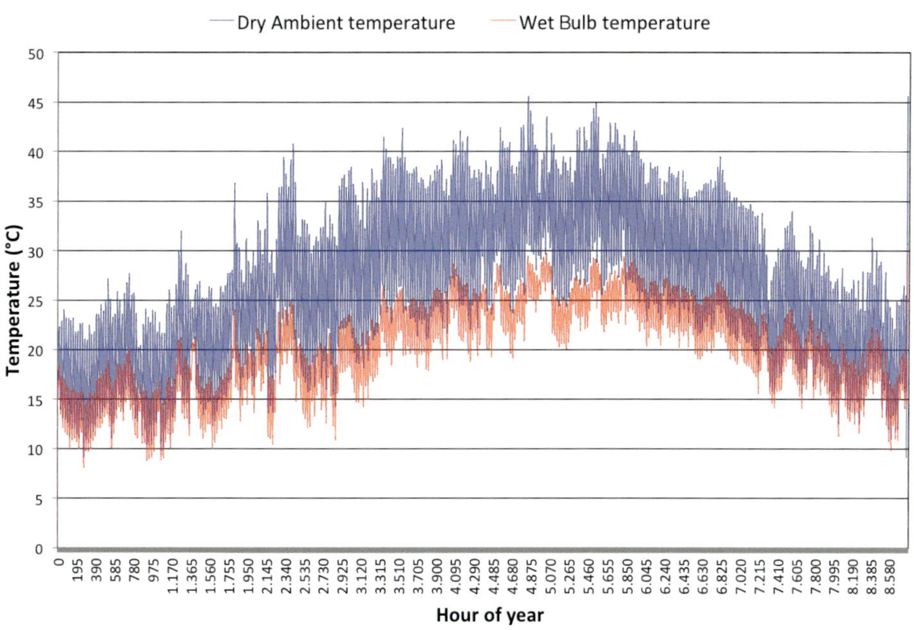

Figure 6.12 Annual values of ambient air (blue) and wet bulb temperature (red) of Abu Dhabi.

Source: Solem Consulting, based on Meteonorm data

Note: the water consumption of a wet cooling tower has to be accounted for. If water is either not available or very expensive at a given project location then a dry cooler may have to be chosen instead, despite its higher cooling water supply temperatures (and lower annual cooling yield of the chiller).

The annual maximum and annual minimum ambient air temperatures, the annual mean maximum and mean minimum temperatures, the average humidity and the annual maximum wet bulb temperatures for all three examples are summarised in Table 6.3.

The radiation data analysis for System No. 2 (Abu Dhabi) is carried out for a tilted collector surface of 24° as design value. This is the collector inclination angle to horizontal for this system.

A parabolic trough collector is chosen here, so only direct normal radiation (DNI) has to be accounted for (see Chapter 2.1.4 for details on parabolic troughs). The analysis shows that the DNI for Abu Dhabi on a 24° tilted surface is 2,125 $kWh_{rad}/(m^2\ a)$. Next, we need to determine the design radiation level for the investigated location site as input for the design of the solar collector field. Figure 6.13 shows the number of hours per year that solar radiation on a 24° tilted surface has a value between two radiation levels. A difference in radiation of 100 W/m^2 in Figure 6.13 is called a radiation bin. The green data are the number of hours per year where radiation occurs between the given limits of a bin. The blue data are the energy content of this radiation bin. This is the number of hours multiplied by the radiation level itself. This analysis is done in order to gain an insight into which bin contains the greatest amount of energy. It can be seen that the most energy received on a 24° tilted surface in Abu Dhabi is contained in the 800–900 W/m^2 bin; hence the design solar radiation for the solar cooling system in Abu Dhabi was chosen as 800 W/m^2. Table 6.4 shows the summary of radiation analysis for all three different investigated locations.

Table 6.3 Weather data for the three investigated system designs

	No. 1 (Munich)	**No. 2 (Abu Dhabi)**	**No. 3 (Singapore)**
Latitude/longitude	48°08'N/11°34'E	24°28'N/54°22'E	1°18'N/103°50'E
Annual maximum ambient air temperature	30.1 °C (86.2 °F)	45.6 °C (114.1 °F)	33.7 °C (92.7 °F)
Annual minimum ambient air temperature	–15.3 °C (4.5 °F)	9.1 °C (48.4 °F)	21.0 °C (69.8 °F)
Annual mean maximum ambient air temperature	16.9 °C (69.4 °F)	33.7 °C (92.7 °F)	28.2 °C (82.8 °F)
Annual mean minimum ambient air temperature	2.3 °C (36.1 °F)	20.0 °C (68.0 °F)	25.9 °C (78.6 °F)
Average ambient air humidity	48.1 per cent	50.4 per cent	81.1 per cent
Annual maximum wet bulb temperature	20.9 °C (69.6 °F)	29.3 °C (84.7 °F)	28.5 °C (83.3 °F)

Source: Meteonorm

Chiller

The design temperature differences between the inlet and outlet of the three external ABsorption chiller circuits (hot, chilled and cooling water) are assumed for Abu Dhabi, as shown in Table 6.5.

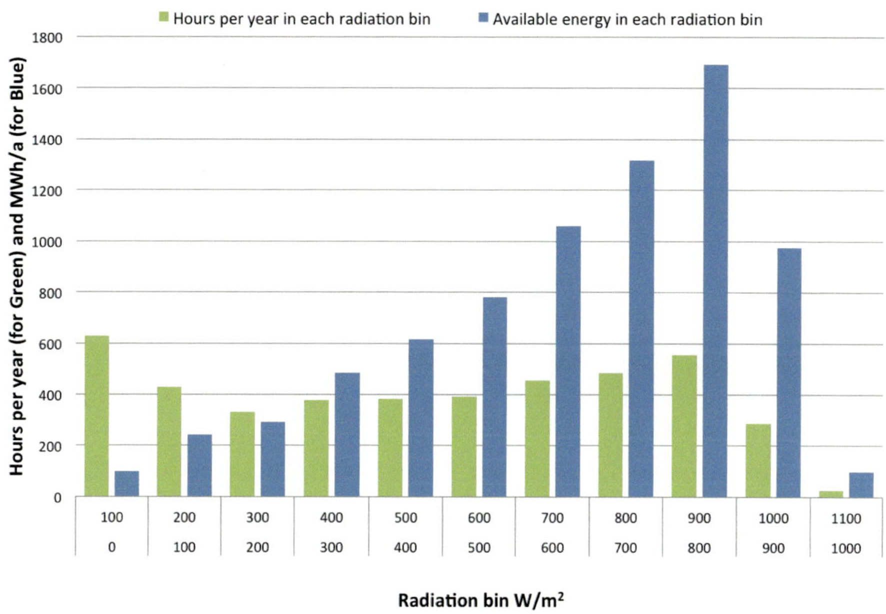

Figure 6.13 Hours per year on a 24° tilted surface (green) and available energy (blue) in each radiation bin; for Abu Dhabi.

Source: Solem Consulting, based on Meteonorm data

Table 6.4 DNI radiation and design solar radiation

	No. 1 (Munich)	No. 2 (Abu Dhabi)	No. 3 (Singapore)
Climate	Moderate climate	Hot climate	Humid climate
Radiation data site (Meteonorm)	Munich	Abu Dhabi	Singapore
Latitude	48.08°	24.28°	1.18°
Annual DNI radiation	1,394 kWh/(m^2 a)	2,125 kWh/(m^2 a)	1,625 kWh/(m^2 a)
Design radiation for manual method	800 W/m^2	800 W/m^2	700 W/m^2

Source: data from Meteonorm

Solar field and heat storage

In general, the solar field can be designed with respect to the peak cooling capacity of the chiller ('sizing to chiller') or to the available roof area on site ('sizing to roof'). The presented examples are all calculated to the peak cooling capacity of the chiller.

All three systems are designed assuming an average design radiation of 700 W/m^2 (Singapore) or 800 W/m^2 (Munich, Abu Dhabi) for each investigated location for a tilted roof installation (see Table 6.4). Different solar thermal collectors are investigated due to the different specifications of the different solar cooling systems (e.g. heating temperatures – see Table 6.5). The annual collector efficiency of the investigated collector tilts is based on the calculated design collector efficiency multiplied by 85 per cent to account for the heat losses from the solar collector circuit (rule of thumb). Table 6.6 shows the preliminary design for the collector fields as well as hot water storage size and time.

Performance check

The annual performance for the solar cooling system at nominal design conditions has been calculated for the example locations of Munich (Germany), Abu Dhabi (UAE) and Singapore, as listed in Table 6.7.

Table 6.5 Design data for the three systems being designed

	No. 1 (Munich)	**No. 2 (Abu Dhabi)**	**No. 3 (Singapore)**
Chiller technology	Single-effect ADsorption	Double-effect ADsorption	Liquid DEC System
Total nominal cooling capacity	10 kW$_r$ (2.8 RT)	34 kW$_r$ (10 RT)	–
Total nominal air volume flow	–	–	10,000 m^3/h
Annual COP	0.6	1.1	1.16
Thermal power required for nominal cooling capacity of chiller	16.7 kW$_{th}$	30.9 kW$_{th}$	52.5 kW$_{th}$
Design hot water supply temperature	72 °C (162 °F)	180 °C (356 °F)	70 °C (158 °F)
Design hot water temperature difference	6 K	16 K	6 K
Design chilled water supply temperature	15 °C (59 °F)	7 °C (44.6 °F)	N.A.
Design chilled water temperature difference	3 K	7 K	N.A.
Heat rejection capacity	26 kW$_{th}$	69 kW$_{th}$	N.A.
Design cooling water supply temperature	27 °C (80.6 °F)	32 °C (89.6 °F)	N.A.
Design cooling water temperature difference	4.5 K	7 K	N.A.

Note:

All values are close to reality, but nevertheless fictitious.

Table 6.6 Calculated solar collector and heat storage data for the three systems

	No. 1 (Munich)	No. 2 (Abu Dhabi)	No. 3 (Singapore)
Collector type	Evacuated tube	Parabolic trough	Flat plate
Heat transfer fluid	Water–glycol	Pressurised water	Water–glycol
Gross area required	43 m^2	94 m^2	144 m^2
Area on roof required for installation	76 m^2	162 m^2	247 m^2
Design collector efficiency	49 per cent	46 per cent	56 per cent
Annual collector efficiency	42 per cent	39 per cent	48 per cent
Design heat storage time for nominal load operation	2.0 h	1.5 h	N.A.
Storage tank size	1.0 m^3	2.5 m^3	1.5 m^3 [1]

Notes

All values are close to reality, but nevertheless fictitious.

[1] Brine storage to store high concentrated salt solution for the dehumidification of the supply air

Table 6.7 Calculated annual performance

	No. 1 (Munich)	No. 2 (Abu Dhabi)	No. 3 (Singapore)
Annual specific solar system yield	0.52 $MWh_{th}/(m^2\ a)$	0.68 $MWh_{th}/(m^2\ a)$	0.70 $MWh_{th}/(m^2\ a)$
Total annual solar system yield	22.5 MWh_{th}/a	64.1 MWh_{th}/a	100.3 MWh_{th}/a
Total annual heat rejected	46.0 MWh_{th}/a	134.6 MWh_{th}/a	N.A.
Total annual cooling provided	13.5 MWh_r/a	70.5 MWh_r/a	116.3 MWh_r/a

Generic hydraulic scheme

A generic hydraulic scheme is shown for System No. 2 (Abu Dhabi), which consists of the solar collectors, the heat storage tank (Figure 6.14) as well as the ABsorption chiller and the heat rejection (wet cooling tower), as shown in Figure 6.15.

Summary

The results of the manual design method presented in Table 6.2 are given in Table 6.8 to illustrate the methodology for System No. 2 (Abu Dhabi) with a double-effect ABsorption chiller. Example values are given here – these would need to be replaced by those with the system-specific values if designing another system using this method.

Note: while fictitious example systems are used here, all numerical values given are based on real components and – where possible – were chosen reasonably close to reality. However, it is not recommended to use these as a

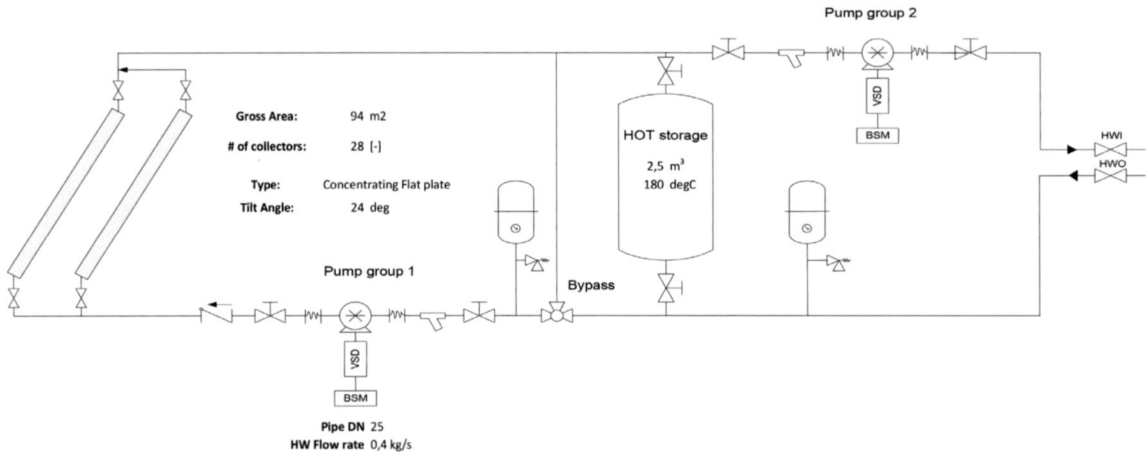

Figure 6.14 Generic hydraulic scheme of System No. 2 (Abu Dhabi), showing the solar collectors, pump and heat storage. Scheme continues in Figure 6.15.

Source: Solem Consulting

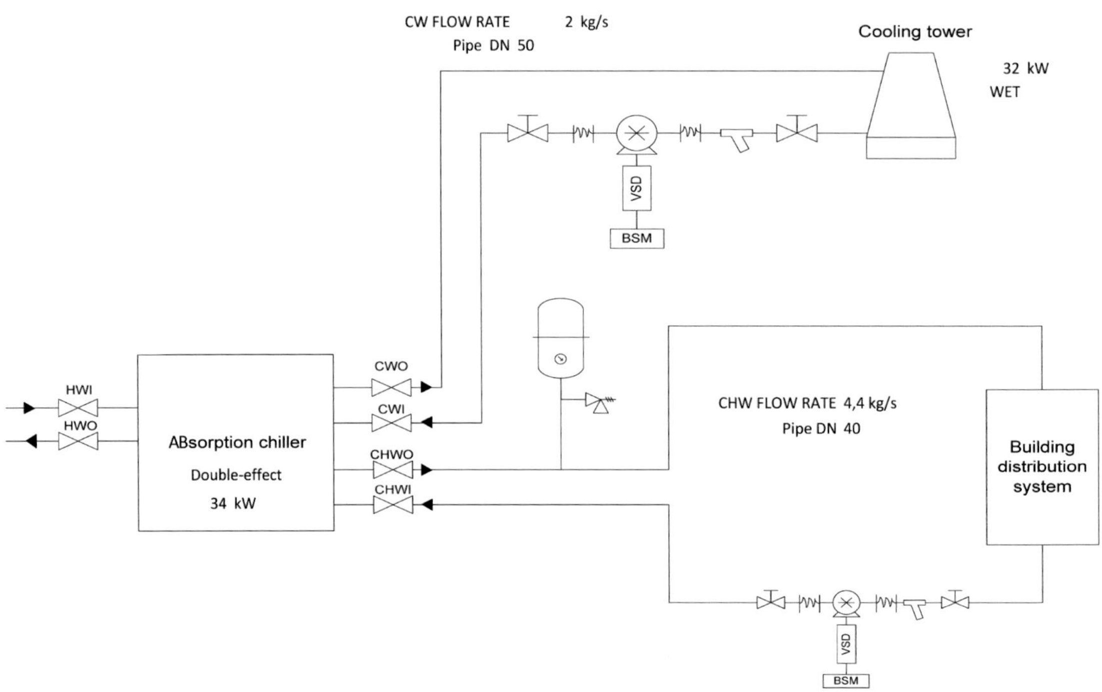

Figure 6.15 Generic hydraulic scheme of System No. 2 (Abu Dhabi), showing the ABsorption chiller, heat rejection (wet cooling tower) and cold distribution. Scheme continued from Figure 6.14.

Source: Solem Consulting

Table 6.8 Manual system design, yield and load parameter for System 2, Abu Dhabi

Group	Parameter	Symbol	Unit	Value[1]
Building	Annual cooling load of building	$Q_{build,c}$	MWh_r/a	70
	Annual heating load of building	$Q_{build,h}$	MWh_{th}/a	0
	Annual hot water load of building	$Q_{build,hw}$	MWh_{th}/a	30
Climate data	Annual DNI radiation on collector surface	$G_{DN,tilted,ann}$	$kWh/m^2/a$	2,125
	Design radiation for manual method	G_{Design}	W/m^2	800
Chiller	Nominal cooling capacity of chiller	$\dot{Q}_c$	kW_r	34
	Average annual COP chiller	$COP_{th,ann}$	–	1.1
	Thermal power required for nominal cooling capacity of chiller	$\dot{Q}_{th}$	kW_{th}	30.9
Heat rejection	Thermal power of heat rejection unit (wet/dry cooler) required for nominal operation of chiller	$\dot{Q}_{CW}$	kW_{th}	69
Solar collectors and heat storage	Annual average collector efficiency	$\eta_{coll,ann}$	–	0.39
	Annual average daytime global irradiation on collector surface	$E_{g,surf}$	W_{rad}/m^2	800
	Annual average global insolation on collector surface	$G_{g,surf}$	$kWh_{rad}/(m^2a)$	2,125
	Collector area required	$A_{coll,req}$	m^2	94
	Collector area oversize	f_{os}	%	0
	Total collector area	$A_{coll,tot}$	m^2	94
	Hot water storage tank volume	$V_{storage}$	m^3	2.6
Control	Control strategy has been implemented	–	–	–
Performance check	Annual thermal energy supplied to chiller	Q_{th}	MWh_{th}/a	78.4
	Annual cooling energy provided by chiller	Q_c	MWh_r/a	70.5
	Full load operation hours per year	n_{op}	h/a	2,059
	Annual heating load of building	$Q_{build,h}$	MWh_{th}/a	0
	Annual hot water load of building	$Q_{build,hw}$	MWh_{th}/a	30
	Thermal energy required for cooling load of building	$Q_{th,req,c}$	MWh_{th}/a	63.6
	Thermal energy required for heating and hot water load of building	$Q_{th,req,hhw}$	MWh_{th}/a	30
	Solar fraction cooling	SF_c	%	100
	Solar fraction heating and hot water	SF_{hhw}	%	49
	Total back-up energy required	Q_{backup}	MWh_{th}/a	15.2
Performance check	Back-up boiler efficiency	η_{backup}	–	0.9
	Total primary back-up energy required	$Q_{backup,PE}$	MWh_{th}/a	16.9
	Solar circuit pump power consumption	P_{solar}	kW_{el}	0.4
	Hot water circuit pump power consumption	P_{hw}	kW_{el}	0.4
	Thermal chiller power consumption	P_c	kW_{el}	0.9
	Cooling water pump power consumption	P_{cw}	kW_{el}	0.8
	Heat rejection unit fan power consumption	P_f	kW_{el}	1.0
	Chilled water pump power consumption	P_{chw}	kW_{el}	0.6
	Control and monitoring power consumption	P_{ctrl}	kW_{el}	0.2
	Total auxiliary power consumption	$P_{aux,tot}$	kW_{el}	4.3
	Electrical COP of total system at nominal operation.	$COP_{el,sys}$	-	7.9

Note

1 Example values are given here – these would need to be replaced with the system-specific values if designing another system using this method.

basis for any designs you may carry out. Readers are required to replace all design values with their own.

The electrical COP, $COP_{el,sys}$ at nominal conditions is a first indicator for energy savings during the design process. It should be greater than the COP of a comparable conventional air-conditioning system. A recommended minimum value for $COP_{el,sys}$ is 8.0. High-quality conventional chillers achieve a $COP_{el,sys}$ of approximately 6.0 at nominal design conditions, but can reach a COP of 10.0 at 60 per cent part-load and even 11.0 at 40 per cent part load [35]. It is therefore important to try to assess the performance of a solar cooling in as much detail as possible. This is best done using design software.

6.4 Design software

In addition to the already presented design methods (Chapters 6.1–6.3), simulation software can be used. This enables a much more accurate system design based on annual performance simulations. Several simulation software tools are available which enable a designer to perform annual system performance simulations at various levels of detail. With these simulation tools, parameters such as solar field size, storage tank size, chiller size, cooling and solar fractions and auxiliaries, etc. can be modelled.

The three most suitable currently available software packages (TRNSYS, INSEL, PolySun) are described below, outlining their capabilities and levels of complexity. TRNSYS and INSEL are most suitable for senior experts from universities, institutes and research companies or highly qualified engineering offices. PolySun is more suitable for beginners or experts.

TRNSYS: Transient System Simulations

TRNSYS, developed at the University of Wisconsin, USA, is a transient system simulation tool with a modular structure in which the user specifies components that constitute the system and the manner in which they are connected (see Figure 6.16). It contains component libraries for both thermal and electrical energy systems. The software tool itself has great flexibility with regard to creating models. It is a strong tool for complex analysis, but the validation of models is essential.

There are component libraries for HVAC equipment, controller components, electrical components, heat exchangers, hydrogen systems, hydraulics, building loads and structures, output devices, physical phenomena, solar collector components, thermal storage components, utility components, weather data reading and processing, STEC components (concentrating solar power) and TESS system components (extended HVAC and solar). These libraries also include AB-/ADsorption, as well as DEC and liquid sorption system models.

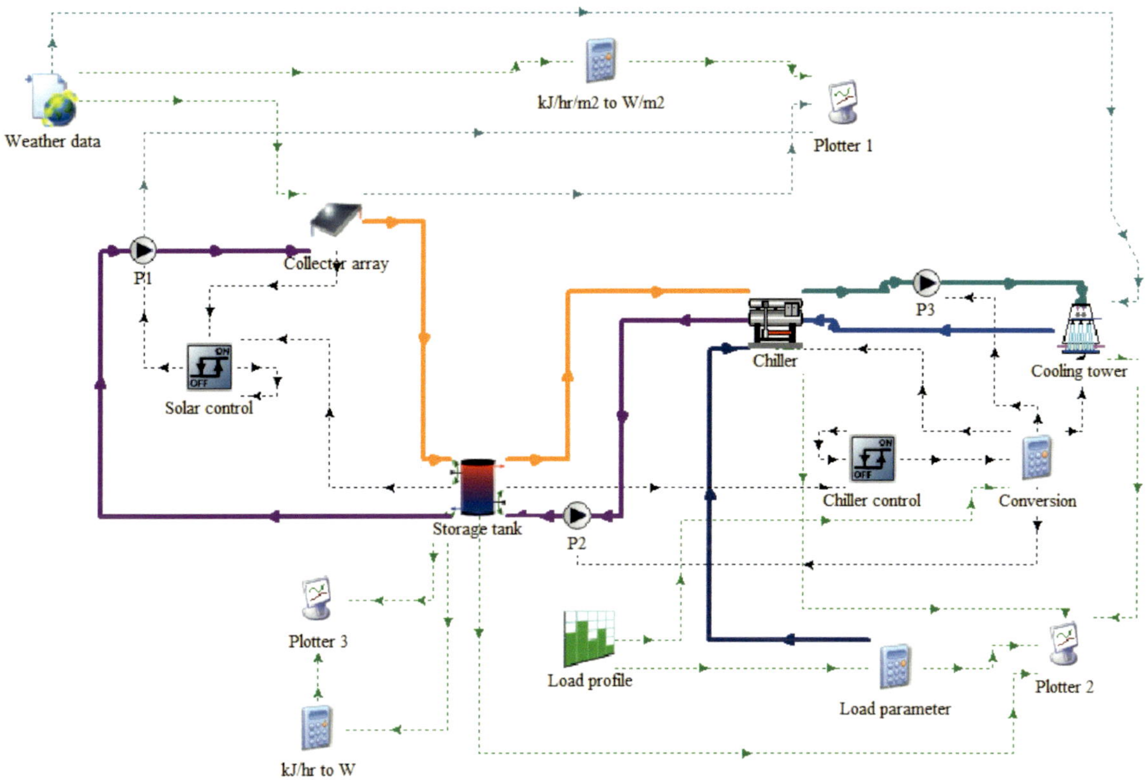

Figure 6.16 Screenshot of TRNSYS simulation environment, showing an example of a simple solar thermal cooling system.

Source: Solem Consulting

INSEL: Integrated Simulation Environment Language

INSEL was developed at the University of Oldenburg, Germany, and later at the Stuttgart University of Applied Sciences, Germany. The structure is similar to TRNSYS (modular, interconnected components). The program HP VEE is used as a graphic editor (see Figure 6.17). There are fewer libraries/toolboxes available than in TRNSYS, but it requires less time/work to create one's own models (using the software FORTRAN as an easy programming language). It is has less flexibility than TRNSYS, but is also a strong tool for complex analysis, especially for control issues. The validation of models (so-called 'blocks') is again essential. The following toolboxes are available: fundamental blocks for numeric and data handling, energy meteorology, solar electrical systems, solar thermal systems, solar thermal power plants and building simulation (under development). Detailed AB-/ADsorption and DEC system blocks are available to simulate solar cooling systems.

Note: all functions in INSEL can be used independently in other software environments, such as Excel, LabVIEW, MATLAB/Simulink, PV-SOL and T-SOL, as well as in one's own software development.

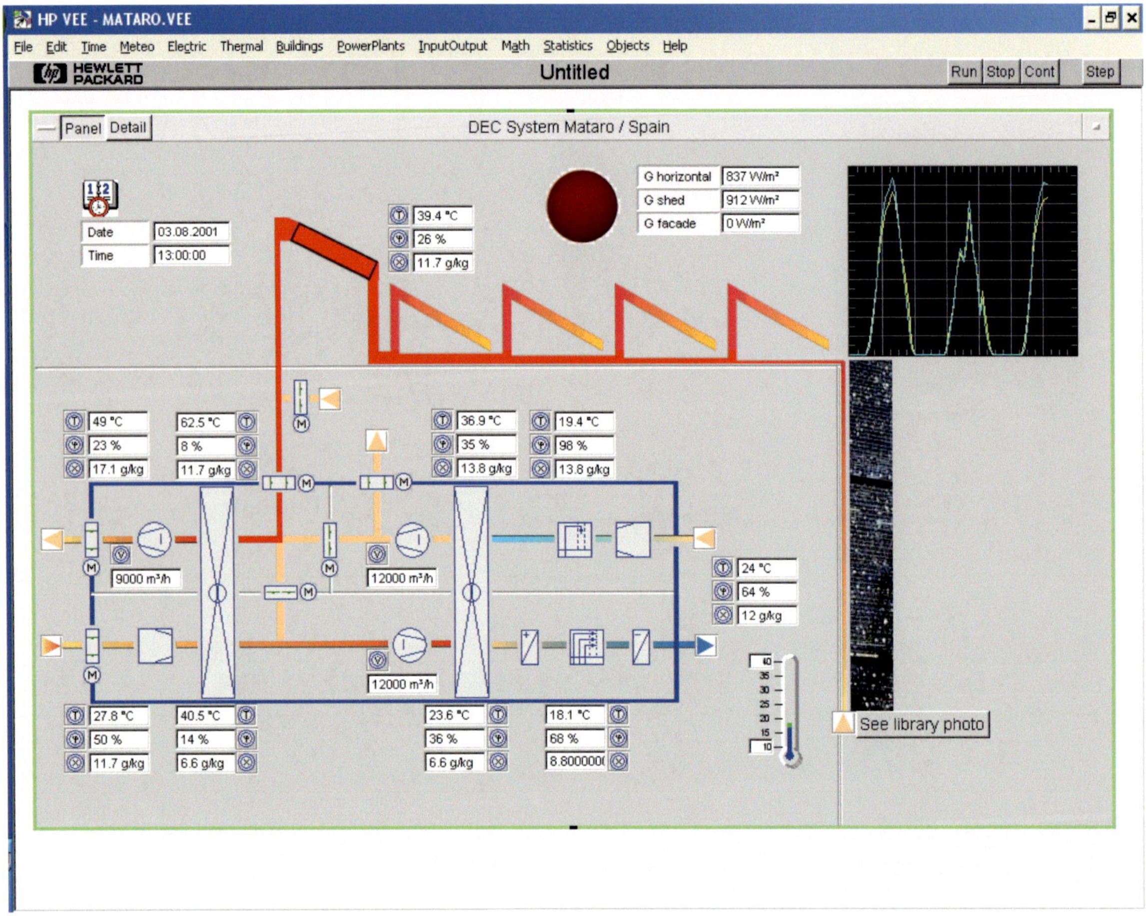

Figure 6.17 Screenshot of INSEL/HP VEE simulation environment showing an example of a solar-assisted DEC system for a library in Spain.

Source: zafh.net

Polysun

The Swiss company Velasolaris, a spin-off of the solar institute SPF in Switzerland, has developed the Polysun simulation software for the design, planning and optimisation of renewable energy systems (see Figure 6.18). The simulation software is modular and component based. There are some libraries and also a simple building model (single zone) included. So far only ABsorption and ADsorption chiller (closed cycles) are included in the libraries to simulate closed-loop solar cooling systems. Open-loop systems cannot be simulated. Polysun is also limited in performance modelling and integration of one's own control algorithms in comparison with TRNSYS and INSEL. The Polysun software tool can be used as a fast pre-design tool for pre-programmed system schemes.

Comparison

TRNSYS is the most powerful simulation tool for solar cooling system design, but also most complex to use. It includes a wide range of built-in component libraries. INSEL also has a good range of libraries, but is not as flexible as TRNSYS; but the graphic user interface (GUI) is more user-friendly compared to TRNSYS. The Polysun tool has a limited range of libraries and control options and is limited to closed sorption chiller cycles only. Its GUI is also very user-friendly, as is its weather database. All three simulation tools are available as free trial versions and can be downloaded from the respective company websites. We recommend trying out each simulation tool and becoming familiar with each before making a purchase. Table 6.9 compares the three presented software tools to identify the capacities and complexity of each.

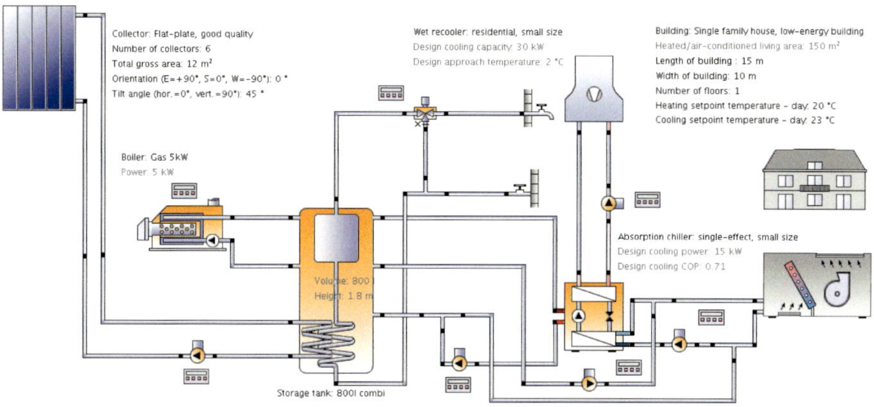

Figure 6.18 Screenshot of Polysun simulation environment showing an example of a solar cooling system with an ABsorption chiller.

Source: Polysun

Table 6.9 Overview of simulation tools

	TRNSYS	INSEL	POLYSUN
HVAC library included	Yes (additional purchase)	Yes	Yes
Building toolbox included	Yes	Under development at time of writing	One zone only
Closed/open cycles possible	Yes	Yes	Closed only
New (own) programmed component models possible to add to the software tool	Yes	Yes	Yes (limited)
Flexibility	High	Medium–High	Medium
Effort required to create a model	High	Medium	Medium
Approximately cost in Europe (single license, excl. VAT)	€5,350	€1,700	€3,300

7

Installation, operation and maintenance

Essential terms for the installation, operation and maintenance of solar cooling systems

DN pipe size – diameter nominal: a measure for pipe diameter according to EN ISO 6708.

Drain-back system – a solar thermal system with a separate heat carrier fluid catchment tank. The heat carrier fluid contained in the collectors and piping above the catchment tank drains into this tank if the solar circuit pump is stopped. This protects the fluid from damage due to frost and stagnation overheating.

NPS pipe size – nominal pipe size: a measure for pipe diameter according to the American National Standards Institute (ANSI).

Solar thermal cooling systems – if properly designed and installed – only require low-level maintenance. This chapter explains how this can be achieved. While a well-designed system looks good on paper, it is really the conversion into a real system that poses the most problems. Malfunctioning systems due to improper installation by unqualified technicians have been observed all too often in the past. An installation undertaken by an appropriately qualified technician will ensure proper operation and long-lasting performance of the system. See Chapter 12.4 for a list of training courses for solar thermal cooling installations.

7.1 Best practice for installations

A diversified study of the performance of more than 120 solar cooling systems was performed as part of the IEA's solar heating and cooling programme from 2006 to 2010 [16]. Multiple recommendations (Do's and Don'ts) for both design and installation were derived from the study. Best-practice recommendations for end-users, investors, building and factory owners planning to install a solar cooling system are given in the following.

Best practice guidelines for the installation of solar cooling systems

- Select one person within your company to be responsible for overseeing and coordinating the whole process of design, installation and commissioning of the system. Make sure communication between all parties involved (manufacturers, technicians, planners and end users) is managed properly.
- Choose an installation company that already has experience with solar thermal and/or cooling systems. First-time installers typically experience problems with the installation and commissioning of systems.
- Check whether pre-engineered solutions (package solutions, solar cooling kits) are available for your project. The use of these typically reduces the time spent on design, installation and commissioning drastically. Further, the labour required for the installation is reduced. Maintenance plans can be included as well in these packages and should be seriously considered. See Chapter 7.4 for details on maintenance.
- Ask the supplier/installation company for an installation schedule that details all different technical installation services involved (plumbing, electrical installation, etc.). Request that exact milestones for each subtask of the project are set. Aim for parallel work of different trades to save installation time.

Source: [16]

7.2 Installation guidelines

This chapter gives an overview of the main installation-related problems that have been experienced with solar cooling systems. Mostly, we will be discussing systems using flat plate or evacuated tube collectors; however, unless otherwise stated, the same sorts of issues can arise with all other liquid fluid-based collector types. Air collectors are discussed separately since they do not use a liquid fluid.

7.2.1 Solar collectors and accessories

Problems related to the position of solar collectors and other components can arise during the installation of solar fields. Components need to be positioned and installed correctly, otherwise system performance and/or safety can be compromised. Common installation mistakes that have been observed include [17]:

- expansion tanks mounted in the wrong place;
- check valves installed the wrong way round;
- temperature sensors mounted in the wrong location or without proper thermal contact;
- piping pressure shocks (steam hammer) occurring due to unfavourable collector orientation;
- insufficient pipe slope in drain-back systems.

Good design practices need to be followed in order to prevent installation mistakes. Appropriately qualified installers should be employed for the system installation. A good compilation of how to avoid installation mistakes in solar fields can be found in literature [17]. Nevertheless, a few critical issues during the installation of key components in solar fields are discussed below.

Temperature sensors

All sensors in a solar system need to measure the values they are supposed to measure. This may sound obvious, but experience with solar systems shows that temperature sensors often read the wrong temperature due to inexpert installation. There are two common ways of installing temperature sensors in solar thermal systems: (1) inserted in the fluid flow via piping sleeves and/or (2) as strap-on sensors on the outside of the piping. Piping sleeve sensors should be inserted against flow direction (preferred), or in such a way that the fluid flows orthogonal (at a right angle) to the sleeve (see Figures 7.1 and 7.2). They are commonly installed at the collector inlet and outlet. Preferably, T-pieces should be used to insert sensor sleeves in this way. The sleeve allows the exchange of a damaged sensor without having to open and drain the fluid piping. This reduces the system downtime.

Strap-on sensors are commonly used at the storage tank, usually on the tank outside wall, or sometimes on the tank piping. Either way, they need to be underneath the insulation, not on top of it. They need to have sufficient thermal contact with the tank wall or the pipe. This can be achieved by using metal hose clamps to affix the sensor to the wall/pipe. They are temperature

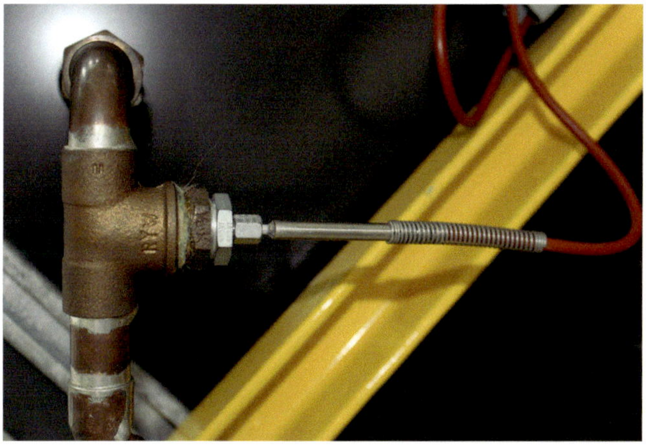

Figure 7.1 T-piece with piping sleeve orthogonal (at a right angle) to fluid flow. Fluid flow is from top to bottom in this example. The sleeve tip should be positioned in the centre of the fluid pipe. Only then is it in good contact with the fluid, hence the temperature measurement will be accurate. If the sleeve tip is too long (i.e. positioned close to the opposite pipe wall or even touching it) then temperature readings will be distorted. On the other hand, if the sleeve is too short (i.e. tip ending within the branching leg of the T-piece) then the tip will not extend into the fluid flow and temperature readings will also be biased. Heat conductive thermal paste should be used within the sleeve to establish good contact between sensor and sleeve.

Source: Solem Consulting

Figure 7.2 T-piece with piping temperature sensor sleeve against fluid flow. Fluid flow is from top to bottom in this example. The sleeve tip should be positioned in the centre of the branching leg of the T-piece. It is not critical if the sleeve extends further above that. A shorter sleeve should be avoided as temperature readings will be biased if the tip does not extend into the flowing part of the fluid. This is the preferred installation for the temperature sensor sleeve because the chances of improper installation are low. As long as the sleeve is long enough to extend into the flow the readings will be accurate. Heat conductive thermal paste should be used within the sleeve to establish good contact between sensor and sleeve.

Source: Solem Consulting

resistant over the operating range of the solar system. The use of plastic cable ties is also possible, but only if the plastic is rated to withstand the maximum system temperatures. Plastic cable ties should not be used in close proximity of the solar collectors. Some strap-on sensors come with copper clips that attach to the pipe. The use of heat-conductive thermal paste is recommended to ensure proper conductivity between tank/pipe surface and sensor.

Expansion vessels

The expansion vessel equalises changes of fluid volume in the system. These changes result from thermal expansion of heat transfer fluid or partial evaporation thereof. Without an expansion vessel, each volume change in the system would result in the pressure relief valve opening and releasing some fluid from the system. Furthermore, the expansion vessel keeps the transfer fluid under a pressure level slightly above atmospheric pressure at all times (back-pressure). If the system is cold and not in operation, the back-pressure (approximately 0.5–1 bar/7.25–14.5 psi gauge pressure) of the expansion vessel prevents air from entering the system through air vents or slightly leaking fittings. Expansion vessels are an essential component of all closed-loop systems and required for all collector types except air collectors.

Expansion vessels need to be rated for the maximum pressure of the system. They can be plain steel vessels or membrane-type vessels. Plain steel vessels usually have an inert gas blanket (e.g. nitrogen) to prevent oxidization of the heat transfer fluid. They are typically used in thermal oil applications (where the heat transfer fluid is oil and not water) and in systems with higher temperatures and pressures (e.g. systems using concentrating collectors). Membrane type expansion vessels separate the heat transfer fluid from a pressurised nitrogen volume by an elastic membrane. They are suitable for lower temperature and water-based systems (typically up to 120 °C/248 °F).

Figure 7.3 shows different modi operandi of a membrane-type expansion vessel with regard to the fill status. Mode 1 is ‘no connection, no operation’: the vessel is disconnected from the system and completely filled with the membrane containing pressurised nitrogen. Mode 2 is 'connected, cold system': the vessel is connected to the system piping and some cold water is pumped into the membrane during filling; this compresses the nitrogen a little and the membrane is pushed back into the vessel. Mode 3 is 'connected, hot system': during operation of a solar thermal system the water expands thermally. The nitrogen volume is compressed further and the membrane is pushed back into the vessel; the system pressure increases. Mode 3 also appears if stagnation occurs in the solar thermal collectors. Then, additional water is pushed into the vessel due to the evaporation of water in the collectors.

The expansion vessel has to be sized correctly for safe system operation. There are two effects that need to be accounted for: the thermal expansion of fluid in all components and the evaporation of fluid in some components. The latter typically occurs in solar thermal collectors only (see Chapter 2.2). The capacity of the expansion vessel should exceed the water volume which is caused by thermal expansion *plus* the total extra water volume generated during stagnation, i.e. the volume displaced by vapour. Stagnation temperature can reach up to 200 °C (392 °F) for flat plate collectors and up to 300 °C (572 °F) for evacuated tube/CPC collectors. Note that concentrating collectors do not go into stagnation because the tracking mechanism points the mirrors away from the sun in case of interrupted fluid flow. The vapour generation in flat plate, evacuated tube and CPC collectors can extend into the connecting piping near the collectors due to heat conduction from the hot collectors. Components in this part of the piping can be exposed to the local boiling temperature via condensing vapour. For example, a valve or an air vent near a collector outlet can reach temperatures much higher than the (liquid) fluid temperature when vapour from the collectors condenses there.

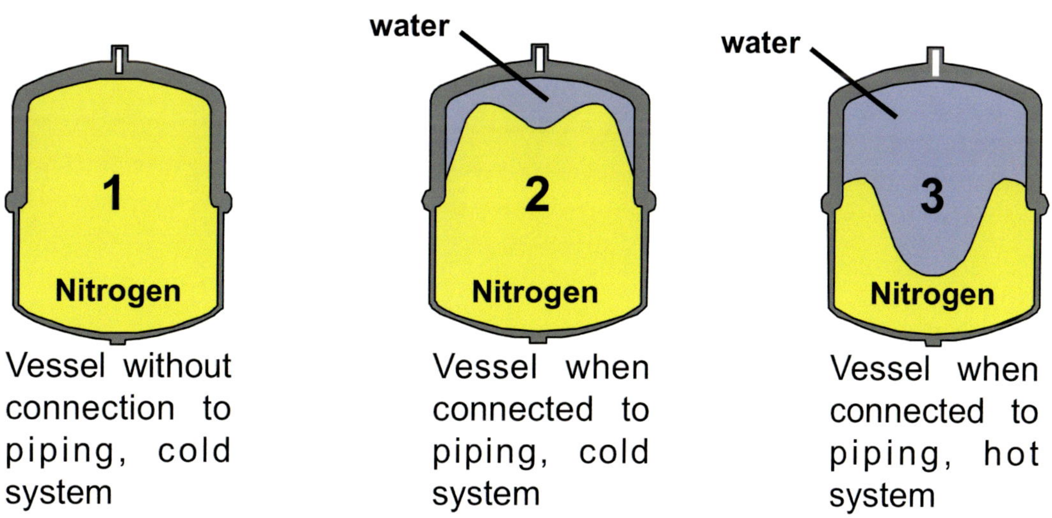

Figure 7.3 Expansion vessel fill status for different modi operandi.

Source: Ritter Energie und Umwelttechnik

The location of the expansion vessel is critical for safe system operation. It should be located on the suction side of the collector circuit pump and on the supply side of the check valve (see Figure 7.4) [18]. This allows the expansion vessel to be filled from both supply and return line at the same time. This reduces the liquid remaining in the collectors during stagnation and equalises the amount of vapour in each line. Consequently, fluid vapour does not penetrate too far into the connecting piping of the collectors. For membrane-type expansion vessels the piping length between collectors and vessel should hold approximately 50 per cent of the expansion vessel capacity. This prevents high-temperature fluid or even vapour entering the expansion vessel during stagnation and damaging the membrane. The expansion vessel should under no circumstances be disconnected from the collector circuit. Therefore protected (capped) valves in the piping between collector and expansion vessel are recommended (see Figure 7.4). For reasons of low pump pressure loss these valves should be full-bore ball valves. These valves do not reduce the internal diameter of the pipe when open.

Check valves

The position of the check valve, as shown in Figure 7.4, is important for the emptying behaviour of the collectors during stagnation. A common mistake is to install the check valve back-to-front, in the wrong flow orientation. The function of the check valve is to avoid reverse fluid flow at night, driven by density differences in the storage tank and the solar collectors. If the collectors cool to ambient temperature at night, the storage tank temperature will most

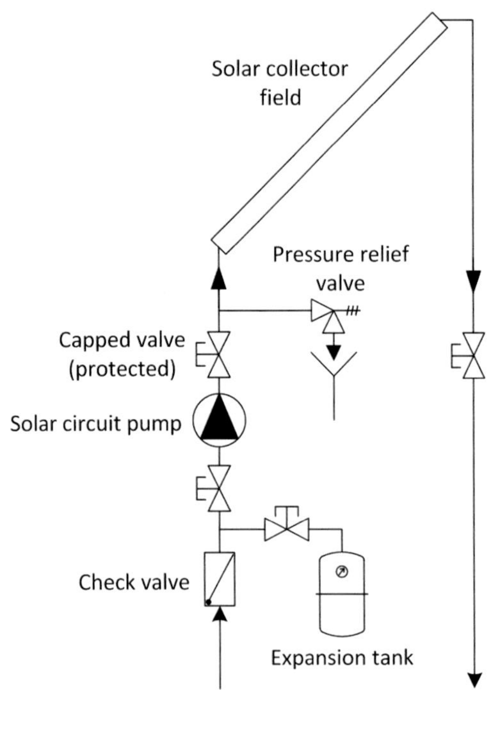

Figure 7.4 Recommended hydraulic arrangement of expansion vessel and check valve.

Source: [17, 18]

likely be higher. In that case, fluid density is higher in the collectors and gravity will start pushing the fluid towards the tank, in the opposite direction to the daytime flow. If the check valve is installed back-to-front in the supply piping, the cold water from the collectors will flow into the tank and displace hotter water in the tank heat exchanger. This displaced water will then rise to the collectors and cool down, thus removing stored heat from the tank. Typically, modern solar installations use pre-assembled component groups, so-called solar stations, where the solar pump, controller, gauges, check valve and flow meter are combined in one unit. This ensures that the check valve is oriented in the correct direction. The use of solar stations is highly recommended, not only from a technical point of view, but also from an economic one. The pre-assembled components save time during installation and significantly reduce the chance of installation errors.

Collector orientation and piping connections

The effect of oscillating and/or condensing vapour during stagnation conditions can cause pressure variations in the system piping. If they occur suddenly, a pressure shock will cause the piping to emit a knocking or banging noise, so-called steam hammer. Steam hammer can be avoided by the correct hydraulic connection of collectors and piping. The correct orientation is to have the inlets/outlets of the collectors at the bottom, not at the top (see Figure 7.5).

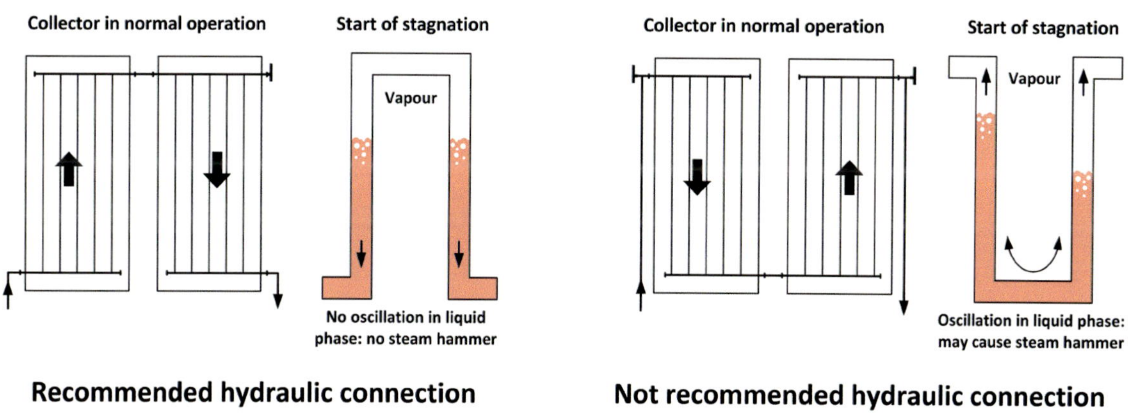

Figure 7.5 Left: recommended hydraulic connection of collectors. Vapour being generated in the collectors will push the liquid fluid simultaneously and relatively quickly out of the bottom of both collectors, and both collectors in this example are emptied at the same time. Right: Unfavourable hydraulic connection of collectors. Fluid is trapped at the bottom of the collectors and cannot be pushed out of the collectors by vapour. The fluid has to evaporate completely. This process does not happen simultaneously and equally in both collectors. The variation in fluid evaporation will result in the liquid columns oscillating, pushing vapour and fluid through the piping. This causes steam hammer. It also takes much longer to empty the collectors of fluid.

Source: [19]

Pipe slope in drain-back systems

A common mistake in drain-back systems is to install the collector circuit piping without sufficient slope. The fluid only drains back automatically into the catch tank when all piping above the catch tank is installed at a downward slope that allows automatic draining due to gravitational forces. Typically, this slope has to be greater than 10 mm per metre of pipe length. Sags in the piping are to be avoided at all times. Note that sags can also be caused during operation by thermal expansion of piping material, especially when long pipe runs are installed. To avoid this, pipe expansion joints (U-shaped pipe sections) should be installed at regular intervals.

7.2.2 Heat/cold storage

Corrosion protection

It is important to install a sacrificial (galvanic) anode in non-stainless steel storage tanks that have continuous mains water intake. Mains water has a high oxygen content, and if no anode is installed, it will cause corrosion within the tank. Sacrificial anodes can be passive or active. Passive anodes are typically made from a magnesia alloy, which has a more negative electrochemical potential than the metal of the tank. The potential difference between tank material and anode material results in corrosion of the anode, not the tank material. Passive anodes require regular checks and replacement when fully consumed. Active anodes use a small electric current that is imposed on the tank to create the potential difference. These are maintenance-free.

Sensor placement

The placement of the storage tank sensor used to read the tank temperature for the solar circuit control is also important – these determine how much water volume is heated to the maximum temperature (see Figure 7.6). If the

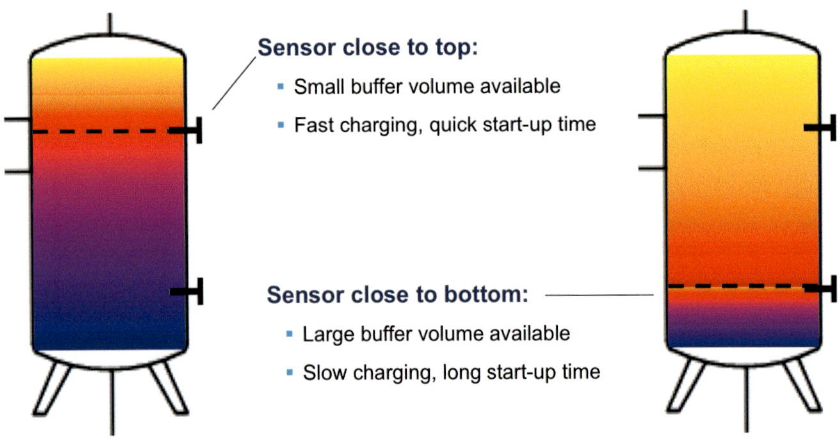

Figure 7.6 Influence of storage tank sensor on charge behaviour.

Source: Solem Consulting

temperature sensor is placed too high on the tank (close to the top), less water is heated, although it is quicker. If the sensor is placed too low, more water is heated, but more slowly. The sensor should have sufficient contact with the tank wall or the port it is assigned to.

Cold storage tank insulation

Cold storage tanks require a special vapour barrier in the insulation to prevent condensation on the tank surface and corresponding piping. Care should be taken to ensure this kind of insulation is applied during installation. Otherwise condensation on the tank surface and piping can cause corrosion or moisture damage to the surroundings.

7.2.3 Chiller

Global delivery of chillers is not a problem; nevertheless, care should be taken when ordering. Points to consider when ordering a thermal chiller are:

- Part-load performance: manufacturers sometimes provide only nominal operational data on the chiller. It is important to ask for the part-load performance curves – solar cooling systems operate in part-load quite often.
- Selecting the right auxiliary energy: some chillers come with a built-in three-way valve for hot water control. This can be driven by compressed air or electrical power.
- Choosing the correct operational voltage/frequency for the country.
- Local building codes: special safety measures apply in most countries for the use of ammonia–water chillers, given the toxicity of ammonia.

7.2.4 Heat rejection

Fan power

A variable speed drive (VSD) attached to the fan motor of the heat rejection unit should be installed. It is one of the most important devices for effective and energy-saving solar cooling system operation (see Chapter 6.3.4).

Frost protection

In frost-prone areas it is mandatory to use a water–glycol mixture in the heat rejection circuit. However, it is common practice for installers to test piping for pressure and leaks using plain water. However, if the heat rejection unit is operated (to test it) at the same time in low ambient temperatures this can cause freezing in the water. Ice crystals forming in the cooler can then damage the pumps. If plain water is left overnight, frost damage can cause piping to burst (see Figure 7.7). It is therefore important to use water–glycol in the commissioning phase if ambient temperatures are low enough to cause freezing.

Figure 7.7 Frost damage of a dry cooler that was operated with plain water in the heat rejection circuit instead of a water–glycol mixture. The second coil from the bottom burst due to ice formation inside the coil.

Source: SolarNext

Water treatment

Wet cooling towers have continuous evaporation of water during operation. Due to the sediment and mineral enrichment of the cooling water, a water treatment plant is recommended. See Chapter 6.3.4 for details.

7.2.5 Other system components

General issues related to the installation of solar cooling systems include the following.

Air vents

Air vents should always be placed at the highest point of the circuit. The installation of multiple air vents is recommended in longer pipe runs, especially if the pipe level changes over its length. Both automatic and manual vents are available.

Automatic air vents automatically release excess air from the pipe but close tight when liquid reaches the vent. In stagnation conditions steam can penetrate into the piping close to the collectors. If steam reaches an automatic vent it can cause it to open. This would result in a loss of heat transfer fluid and system pressure through the vent. Therefore, automatic air vents should not be used in close proximity to the solar collectors (where steam is likely to occur), but they can be used throughout the rest of the system, They need to be rated for the maximum operating temperature of the circuit they are to be installed in.

Manual air vents only release air when opened manually. These should be used in close proximity to the solar collectors. They need to be rated for the stagnation temperature of the solar collectors. Manual vents should be opened at frequent intervals during system commissioning in order to release excess air, and they should be regularly checked at maintenance intervals (see Table 7.1).

Air vents

The correct installation of air vents is important. Excess air in a circuit will cause an air lock at the highest point of the circuit. This can result in stopping fluid circulation even though the circuit pump is operating. In circuits with hot fluid this can damage the pump through overheating and consequent bearing damage. Excess air in water–glycol mixtures results in chemical oxidation reactions causing the pH value to drop. Increased corrosion, especially of soldered parts (pipe connections), can occur, resulting in deposits of solids in the pipe. This can even lead to complete pipe blockage.

Piping insulation

The piping insulation in solar cooling systems needs to be chosen according to the following criteria:

- the highest system temperature in piping (including stagnation temperature of solar collectors);
- indoor/outdoor use;
- appropriate thermal resistance/heat conductivity.

Solar collectors can achieve stagnation temperatures up to 300 °C (572 °F). These high temperatures can be conducted into the connecting piping, even during standstill. The pipe insulation in the solar collector circuit should be temperature-rated up to the stagnation temperature of the collectors.

Some insulation is not UV-rated for outdoor use and will deteriorate rapidly if used externally. Sheet-metal cladding around a non UV-rated insulation material is an option. The insulation used outdoors should be closed-cell foam with sheet-metal cladding. Foam-based insulation (if used without cladding) is a preferred material for nest building by certain animals (birds, mice). Open-cell foam insulation materials are not suitable for outdoor use; they lose thermal resistance when soaked with water, and will not recover it until the pipe heat has evaporated the water. Open-cell foam can be used indoors if no moisture is present.

The longer a pipe-run is, the more heat is lost along its length. Insulation materials of sufficiently high thermal resistance (i.e. low thermal conductivity) should be chosen. A rule of thumb is to use an insulation thickness of 100 per cent of the nominal pipe diameter up to a pipe diameter of DN 100 (NPS 4").[1] For pipe diameters greater than DN 100, 100 mm-thick insulation is recommended, regardless of the pipe diameter.

Care has to be taken in thermal oil and pressurised hot water systems with regard to the dryness of the insulation. Leaks of rain water or dew into the insulation should be avoided. Also, the leak tightness of pipe fittings is important. A slow-leaking fitting releases fluid into the surrounding insulation

[1] This rule of thumb for insulation thickness has been given for a thermal conductivity of 0.035 W/(mK)/0.012 Btu/(hrftF) of the material. If the conductivity of the material is higher than this, less insulation can be used.

material (usually mineral or glass wool). Water or heat transfer fluid that has soaked into the insulation increases the thermal losses of a pipe by a factor of 3, compared to dry material. This is because the heat conductivity of the fluid is greater than of the (dry) insulation material, thus fluid-soaked insulation material has a lower thermal resistance. One solution is to use closed-cell foam instead of mineral or glass wool. Figures 7.8 and 7.9 show an infrared and the corresponding visible light image of a pipe, insulated with rock wool and sheet metal cladding. Pressurised water of 180 °C (356 °F) is flowing through the pipe and has leaked into the insulation of a 90° pipe bend. Capillary forces have caused the water to spread, resulting in surface temperatures of the cladding above 90 °C (194 °F). This causes additional heat loss of the fluid and also poses a safety concern for operating staff (potentially harmful temperature when touched).

Another possible solution, at least regarding valves, is to install valves in an oil-filled pipe with the handle pointing upside-down. Thus, any leaking oil from the valve stem will drip away from the insulation onto the ground. However, some building codes require special care with thermal oil being released on flat roofs. A leak in the oil piping can potentially result in thermal oil being released into the rain-water collection system of the roof. From there, the oil will contaminate the waste water. A solution is to use a device for containment below the collectors and piping which catches any spills and collects them separately from the rain-water drains. Note: It is important to follow local codes regarding all aspects of a solar cooling system installation. Selected codes are listed in Chapter 12.2.

Pumps

Care has to be taken when using standard hot water/HVAC pumps in hot ambient conditions. Most standard pumps are rated for a maximum ambient temperature of 40–50 °C (104–122 °F). If the pump is located outdoors in a hot climate then the maximum operating pump temperature can be easily reached. Indoor pumps can also reach the maximum temperatures when installed in a boiler room that is (1) exposed to sun, (2) improperly vented or (3) contains multiple heat sources (e.g. a heat storage tank, piping, etc.).

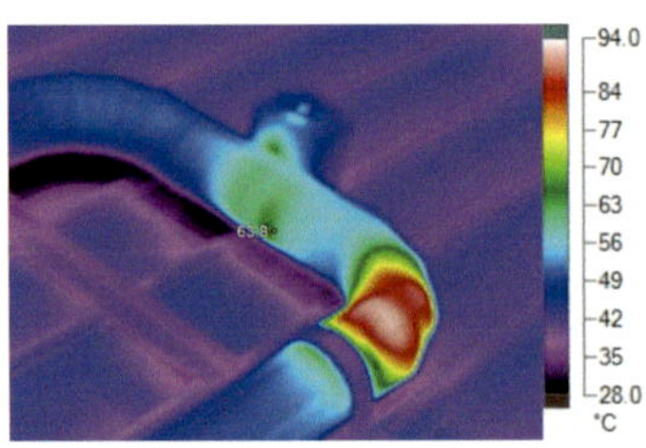

Figure 7.8 Infrared image of water-soaked pipe bend insulation. Surface temperature exceeds 90 °C (194 °F) in sections. The water temperature inside the pipe is 180 °C (356 °F).

Photo: Jeremy Osborne

Figure 7.9 Visible light image of the same pipe bend.

Photo: Jeremy Osborne

7.3 Operation

The following recommendations ensure proper operation of a solar cooling system

- Request an end-user manual from your supplier that details the essential information for the operation of the system. This should include settings for different operational conditions, monitoring, maintenance and control of the system. Also, the contact details of all component suppliers should be included.
- Request the chiller manual from the chiller manufacturer in your language or a language you can understand. This may sound obvious, but not all chiller manufacturers provide manuals in different languages.
- Monitor the performance of your solar cooling system to ensure continuous operation according to the design conditions, ensure energy efficiency and detect possible faults immediately. Most system controllers today allow for remote monitoring and suppliers usually offer analysis of the recorded data as part of a maintenance plan.

7.4 Maintenance

The following points are recommended for maintenance of a solar cooling system

- Select a maintenance company before or at commissioning of the system.
- Ask the supplier/installation company for a maintenance plan, listing required maintenance tasks including time intervals between tasks, plus an analysis of data collected during maintenance.
- Typical maintenance work and schedules are given in Table 7.1; however, additional maintenance or different schedules may be required for specific systems.
- Thermal chillers need to undergo regular maintenance according to the manufacturer's specifications. These vary from model to model and cannot be detailed here. The vacuum in AB- and ADsorption chillers needs to be maintained at all times to ensure proper operation of these chillers. The vacuum should be checked at regular intervals.

Table 7.1 Recommended maintenance actions for solar cooling systems

Time interval	**Maintenance action**
Once per week	Check whether main system parameters are okay (use automatic monitoring where possible) Check for system alarms or warnings (use automatic monitoring where possible)
Once per month	Visual inspection of solar collectors and cleaning if necessary
Once every half year	Check chiller vacuum Clear heat rejection system of solid deposits Clean air intake and outlets of heat rejection system Empty heat rejection system for winter if necessary
Once per year	Check system pressure Check quality of heat transfer fluids (take samples and measure composition) Check pumps and valves for correct operation Clean inline filters Check manual air vents for accumulated air and release if necessary

8 Economic feasibility

This chapter discusses the economic feasibility of solar thermal cooling systems. We present a method to evaluate whether a solar thermal cooling system has a lower or higher lifetime cost than a standard conventional air-conditioning system of similar size. Specific cost breakdowns of selected solar thermal cooling systems are also presented.

Note: the economics of a PV-based air-conditioning system are not discussed in detail here. They depend largely on the subsidy scheme for PV power and the cost for PV panels (see Chapter 3.5). However, the economic calculation method presented in this chapter can also be applied to a PV-based air-conditioning system if the inputs are changed accordingly.

Essential terms for the economic calculation of solar thermal cooling systems

Annuity method – method for investment calculations using the annuity factor. It refers to a time series of constant cash flows over multiple years. Annuity is calculated based on the first year, assuming constant cash flows for all following years

Discount rate – the rate of interest used in NPV calculations. It reflects the rate of return which could be expected on the global financial markets for an investment of similar risk

NPV method – method for investment calculations using net present value (NPV). It refers to a time series of variable cash flows, i.e. changing cash inflows and outflows over multiple years. NPV is the present value of the difference between cash inflows and cash outflows, taking inflation and returns into account.

A solar thermal cooling system is usually considered as an alternative to conventional air-conditioning systems. Its economic viability depends on a variety of parameters, including but not limited to:

- the price for electrical power and/or natural gas;

- the cooling load pattern of the building;
- the solar resource/annual solar radiation;
- local equipment and installation costs; and
- potential subsidies at the project location.

The evaluation of the economic feasibility of a solar thermal cooling system includes both a detailed technical design and a detailed economic analysis. Technical design and economic viability are interlocked; changing one will have effects on the other. If this is the case, iterative calculations will have to be performed. For example, a system designed to use expensive and high-quality solar collectors may have a very good performance but may not meet economic expectations. A solution might be to replace the expensive high-quality collectors with less costly ones and thus meet the economic requirements. However, the cheaper collectors may not have the same efficiency as the replaced ones, thus the number of collectors (the total collector area) has to be increased accordingly – and the economic calculations will have to be performed again. This means that several possible scenarios need to be tried out in order to find the optimum combination of technical design and economic viability.

Double benefit – solar heating and cooling

The solar collectors of a solar thermal cooling system deliver heat whenever sufficient insolation is available. This heat can be used to operate a thermal sorption chiller or desiccant-evaporative cooling (DEC) system to provide cooling. It can also be used for space heating, domestic hot water generation, pool heating and process heat. The calculation of economics for solar cooling systems often focuses on the cooling only. This is not recommended if the use of heat is also possible in a potential project. The double benefit of providing cold and heat is an advantage of solar thermal cooling systems that should not be neglected in an economic calculation. The economics of a cooling-only system will most likely be less favourable than of a combined cooling and heating system.

At first, the solar cooling system needs to be designed to meet the requirements of a given air-conditioning task (see Chapter 6). Then, in order to assess the economic feasibility of the proposed solar cooling system it needs to be compared to a standard conventional air-conditioning or chilled water system using vapour-compression chillers and fan coils or air-handling units. This standard system, of course, has to meet the requirements of the given air-conditioning or cooling task. The economic feasibility of the solar thermal cooling system can then be estimated compared to this standard system. Consequently, this requires the design of a standard system in order to be able to calculate its cost. This design should include the main components (cost drivers), but there is no need to perform a detailed technical design or annual simulations.

A choice of solar cooling design methods, including rules of thumb, checklists and manual calculations, has been described in Chapter 6. While the manual method provides relatively accurate numbers for the component sizing, it does not estimate annual performance. However, in order to assess the economic feasibility of a system it is necessary to also know the following:

- annual yield of solar thermal collector field (this depends on the solar radiation);
- annual cooling energy provided by the system;
- annual back-up heat required (if a gas/oil boiler is used as a back-up);
- annual back-up cold required (if a vapour-compression chiller is used as a back-up);
- annual auxiliary power consumption;
- number of hours of system operation per year.

These data are best and most accurately arrived at or evaluated using annual performance modelling software. Chapter 6.4 gives an overview of available software for this. Modelling the annual system performance using software will give results as close as possible to reality: day-by-day, hour-by-hour or even minute-by-minute system performance is simulated, using location-specific weather data (see Chapter 12.6), accurate load data and the technical specifications of the system components. Literature sources are available for software modelling of solar thermal cooling systems (see Chapter 12.3).

If no software is available manual methods can be used. These use average annual values derived from literature, empirical data or engineering experience. Such a method is described in Chapter 6.3. The significant parameters for economic feasibility can thus either be derived from software modelling or manual calculation.

8.1 NPV method: a worked example

The methodology is presented using a worked example of a closed system using concentrating parabolic collectors and a thermal chiller, providing chilled and hot water loads (see Figure 8.2). This solar thermal cooling system is compared to a standard conventional non-solar system, consisting of a screw chiller providing chilled water and a condensing gas boiler for hot water provision (see Figure 8.3).

Note: the methodology can also be applied to an open DEC system, in which case the chiller, cooling tower and fluid piping have to be exchanged with the DEC unit, heat exchangers and air ducts, respectively.

While fictitious example systems are used here, all numerical values given are based on real components and – where possible – were chosen reasonably close to reality. However, it is not recommended to use these as a basis for any designs you may carry out. Readers are required to replace all design and cost values with their own.

The calculation method for this economic comparison is based on the NPV approach. This approach calculates all cost discounted to the year of installation, assuming a system lifetime and discount rate. The NPV method takes into account changes in the inflation rate and energy prices over the lifetime of the system (although these are hard to predict). Also, capital investments, e.g. repair or replacement costs, are taken into account. The degradation of component performance due to ageing can also be included. The NPV method is a standard method for calculating the economic feasibility of projects and shall not be detailed in this book.

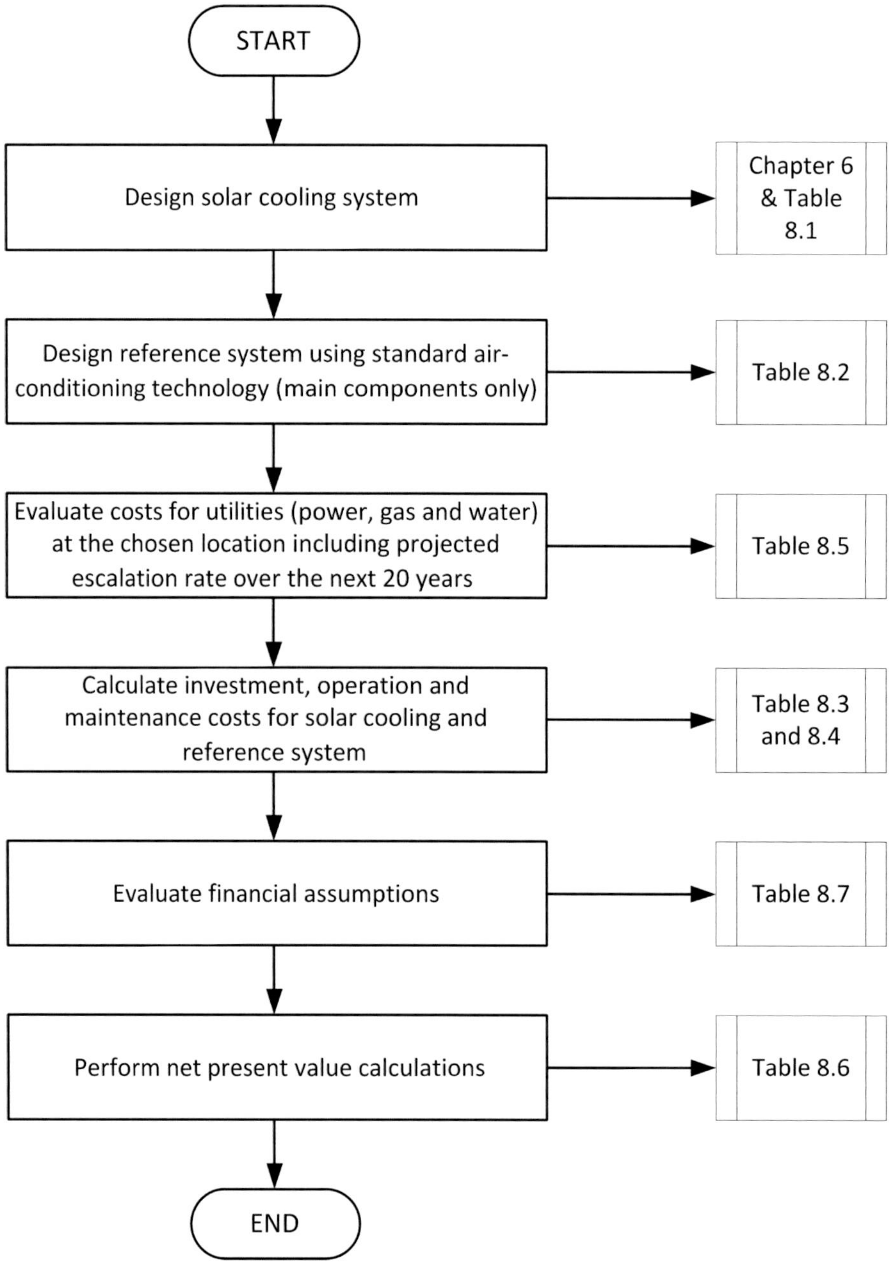

Figure 8.1 Economic feasibility calculation methodology flow chart. This diagram shows the method and steps required to utilise the net present value method. Each individual step needs to be completed in the order shown.

Source: Solem Consulting

The tables given below can be replicated in a spreadsheet program in order to implement the calculation procedure illustrated.

8.1.1 System definition and technical parameters

Both the solar thermal cooling and the reference conventional cooling system need to be designed first. The design needs to be determined from a system design procedure as detailed in Chapter 6.3 or similar. The design of the reference system (see Figure 8.3) should include the main components (cost drivers), but there is no need to perform a detailed annual performance simulation.

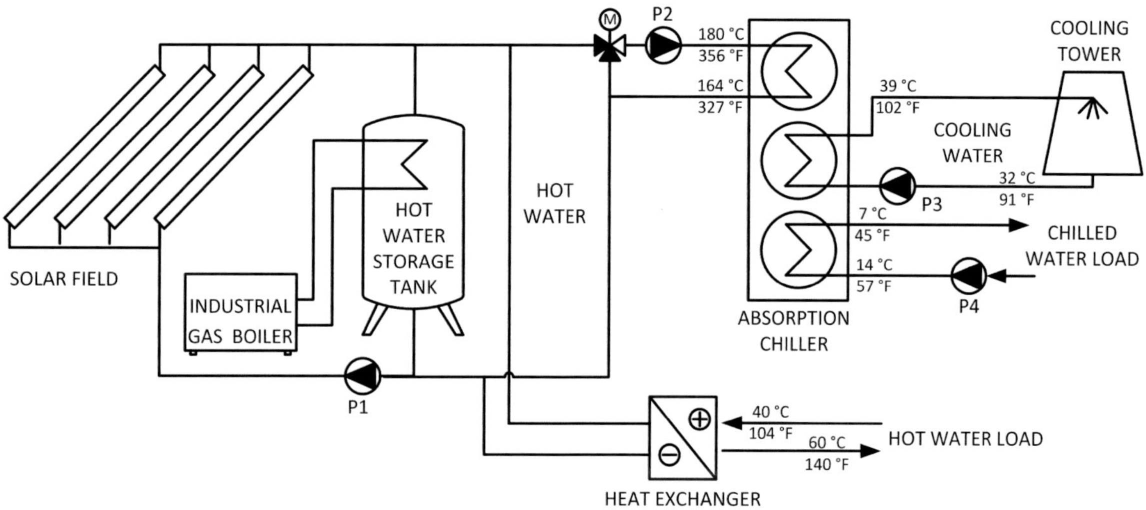

Figure 8.2 Design example of a solar cooling system for the economic feasibility calculation used in this chapter.

Source: Solem Consulting

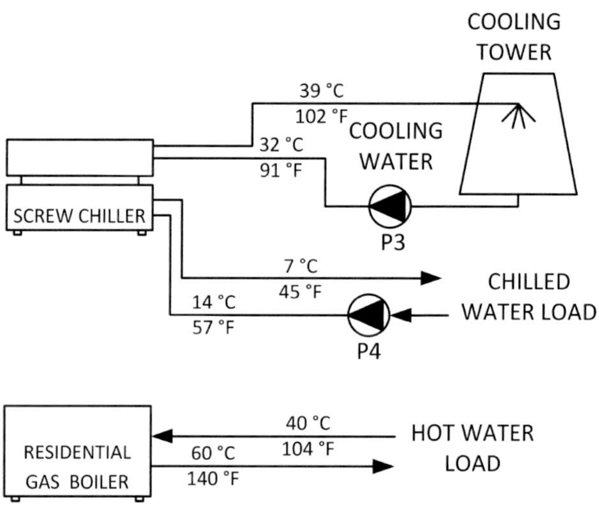

Figure 8.3 Design example of reference conventional cooling system for the economic feasibility calculation used in this chapter.

Source: Solem Consulting

In our example, we assume the design for the two systems is as shown in Figures 8.2 and 8.3. The solar cooling system consists of the components listed in Table 8.1 and provides hot and chilled water loads for a small commercial building of approximately 300 m^2 floor area in Abu Dhabi, UAE. Table 8.1 shows the solar cooling system design parameters. The reference conventional (non-solar) system consists of the components listed in Table 8.2; it also provides chilled and hot water for the same building.

8.1.2 System costs

The next step is to assign costs to all components as well as to system operation and maintenance (O&M). For the investment costs, component quotes should be obtained from manufacturers. Operational costs can be calculated from the expected energy and water consumption of the system and utility costs. The latter can be obtained from local suppliers; growth rates are usually subject to individual estimates. Maintenance costs can either be obtained from manufacturers offering maintenance plans for their components, via extrapolation from known systems or via a percentage of the total investment cost. Tables 8.3 and 8.4 show the assumed investments costs (including installation cost) for solar cooling and conventional reference systems, respectively. We have differentiated between the main components and the balance of plant. Table 8.5 shows the O&M costs for both systems in the worked example.

In Table 8.5 the O&M cost is shown for the first year of operation (year 1). The operational cost is subject to the growth rate of utilities and the maintenance cost subject to the inflation rate (CPI). Both of these factors are considered in the NPV calculation for the systems (see Table 8.6).

Note: the NPV calculations presented here do not include potential salvage values or disposal costs for either system at the end of their working lives. The salvage value of solar thermal cooling systems is mainly determined by the scrap value of the raw materials used. These include aluminium and copper from the solar thermal collectors, including piping, as well as plain/stainless steel from the AB-/ADsorption chillers. Disposal costs for solar thermal cooling systems are mainly determined by the cost for safely disposing of the chemicals used for the different sorption processes. These chemicals include LiBr for ABsorption chillers and LiCl, triethylene glycol ($C_6H_{14}O_4$) and aqueous calcium chloride ($CaCl_2$) for DEC systems.

Table 8.1 Design parameters of example solar cooling system for economic feasibility calculations

	Solar cooling system	Unit	Quantity
Solar field and hot water circuit	Solar collector aperture area	m^2	94
	Thermal power (nominal[1])	kW_{th}	35
	Collector efficiency[2]	%	46
	Annual solar gain	MWh_{th}/a	78
	Collector tilt	° from horizontal	20
	Azimuth	° from south	49 West
	Solar field flow rate (nominal)	m^3/hr	2.0
	Location	–	Abu Dhabi, UAE
	Hot water storage tank size	m^3	2.5
Thermal chiller and chiller circuits	Chiller cooling capacity (nominal)	kW_r	34
	COP (nominal)	–	1.1
	Driving energy (nominal)	kW_{th}	30.9
	Hot water temperature	°C (°F)	In 180 (356) – Out 164 (327)
	Hot water flow rate	m^3/hr	1.7
	Chilled water temperature	°C (°F)	In 14 (57) – Out 7 (45)
	Chilled water flow rate	m^3/hr	4.2
	Cooling water temperature	°C (°F)	In 39 (102) – Out 32 (91)
	Cooling water flow rate	m^3/hr	8.0
	Cooling tower capacity (nominal)	kW_{th}	65
	Cooling tower type	–	Wet, closed loop
	Cooling tower fan control	–	Variable speed
	Cooling tower water consumption (nominal)	m^3/hr	0.1
Loads	Annual chilled water load	MWh_r/a	70
	Annual hot water load	MWh_{th}/a	30
	Full-load hours of operation	h/a	2,059
Auxiliary power consumption	Solar circuit pump	kW_{el}	0.4
	Hot water circuit pump	kW_{el}	0.4
	Thermal chiller	kW_{el}	0.9
	Cooling water pump	kW_{el}	0.8
	Cooling tower fan	kW_{el}	1.0
	Chilled water pump	kW_{el}	0.6
	Control and monitoring	kW_{el}	0.2
Back-up	Net back-up heat required[3]	MWh_{th}/a	15
	Back-up heater efficiency	%	90
	Gross backup heat required	MWh_{th}/a	17
	Backup type	–	Condensing gas boiler
	Backup heater capacity	kW_{th}	50

Notes:

[1] Nominal conditions are 800 W/m² solar irradiation and operation of system at full capacity.

[2] Collector efficiency is given at 800 W/m² solar irradiation.

[3] Net backup heat is heat supplied to the storage tank, gross back up heat is energy supplied to backup heater. The difference between net and gross is the difference between primary (gas) and secondary (heat) energy.

Table 8.2 Design parameters of example reference system for economic feasibilty calculations

	Reference system	Unit	Quantity
Conventional chiller and chiller circuits	Chiller cooling capacity (nominal)	kW_r	35
	COP (nominal)	-	3.0
	Driving energy (nominal)	kW_{el}	11.7
	Chilled water temperature	°C (°F)	In 14 (57) – Out 7 (45)
	Chilled water flow rate	m^3/hr	4.2
	Cooling water temperature	°C (°F)	In 39 (102) – Out 32 (91)
	Cooling water flow rate	m^3/hr	5.5
	Cooling tower capacity (nominal)	kW_{th}	45
	Type	–	Wet, closed loop
	Water consumption (nominal)	m^3/hr	0.05
	Fan control	–	Variable speed
Loads	Annual chilled water load	MWh_{th}/a	70
	Annual hot water load	MWh_{th}/a	30
	Full-load hours of system operation	h/a	2,059
Auxiliary power and water consumption	Cooling water pump	kW_{el}	0.8
	Cooling tower fan	kW_{el}	1.0
	Chilled water pump	kW_{el}	0.6
	Control and monitoring	kW_{el}	0.2
Back-up	Net back-up heat required	MWh_{th}/a	15
	Back-up heater efficiency	%	90
	Gross backup heat required	MWh_{th}/a	17
	Back-up type	-	Condensing gas boiler
	Back-up heater capacity	kW_{th}	50

Table 8.3 Investment costs/parameters of solar cooling system (in US dollars)

	Solar cooling system	Unit	Type/size	Quantity	Specific cost ($/unit)	Main component cost	Balance of plant cost
Solar field and hot water circuit	Collectors	m^2	Concentrating parabolic	94	$650	$60,955	
	Piping	m	DN 32	30	$45		$1,350
	Piping	m	DN 25	20	$35		$700
	Piping	m	DN 20	20	$30		$600
	Pipe lagging	m	DN 32	30	$45		$1,350
	Pipe lagging	m	DN 25	20	$35		$700
	Pipe lagging	m	DN 20	20	$30		$600
	Solar circuit pump	m^3/hr	2	1	$1,500		$1,500
	Valves miscellaneous	–		1	$200		$200
	Heat exchanger	–	Plate	1	$800		$800
	Expansion tank and safety relief valve	–		1	$150		$150
	Hot water circuit pump	m^3/hr	2.4	1	$1,500		$1,500
	Valves miscellaneous	–		1	$200		$200
	Expansion tank and safety relief valve	–		1	$150		$150
	Hot water storage tank incl. insulation	m^3	2.5	1	$10,000	$10,000	
	Piping	m	DN 32	20	$45		$900
	Pipe lagging	m	DN 32	20	$45		$900
Back-up	Condensing gas boiler	kWth	50	1	$8,500	$8,500	

	Solar cooling system	Unit	Type/size	Quantity	Specific cost ($/unit)	Main component cost	Balance of plant cost
Thermal chiller and chiller circuits	Absorption chiller	kWr	34	1	$45,000	$45,000	
	Cooling tower	kW_{th}	65	1	$15,000	$15,000	
	Cooling water pump	m^3/hr	8	1	$1,500		$1,500
	Valves miscellaneous	–		1	$200		$200
	Piping	m	DN 40	50	$50		$2,500
	Pipe lagging	m	DN 40	50	$50		$2,500
	Chilled water pump	m^3/hr	4.2	1	$1,500		$1,500
	Valves miscellaneous	–		1	$200		$200
	Expansion tank and safety relief valve	–		1	$150		$150
	Piping	m	DN 32	50	$45		$2,250
	Pipe lagging	m	DN 32	50	$45		$2,250
Instruments and control	Sensors/gauges	–	Temp/ radiation/ pressure	10	$100		$1,000
	Controller/system PLC	–	PLC	1	$5,000		$5,000
	Total cost (main component/BoP)					$139,455	$30,650
	Percentage (main component/BoP)					82 per cent	18 per cent
	Total cost system					$170,105	

Note

All cost numbers include installation.

Table 8.4 Investment costs/parameters of reference system (in US dollars)

	Reference system	Unit	Type/size	Quantity	Specific cost ($/unit)	Main component cost	Balance of plant cost
Gas boiler	Condensing gas boiler	kW_{th}	50	1	$8.500	$8,500	
Conventional chiller and chiller circuits	Vapour compression chiller, screw type	kW_r	34	1	$45,000	$45,000	
	Cooling tower	kW_{th}	45	1	$9,000	$9,000	
	Cooling water pump	m^3/hr	5.5	1	$1,500		$1,500
	Valves miscellaneous	–		1	$200		$200
	Piping	m	DN 32	50	$45		$2,250
	Pipe lagging	m	DN 32	50	$45		$2,250
	Chilled water pump	m^3/hr	4.2	1	$1,500		$1,500
	Valves miscellaneous	–		1	$200		$200
	Expansion tank and safety relief valve	–		1	$150		$150
	Piping	m	DN 32	50	$45		$2,250
	Pipe lagging	m	DN 32	50	$45		$2,250
Instruments and control	Sensors/gauges	–	Temp/ pressure	5	$100		$500
	Controller/system PLC	–	PLC	1	$5,000		$5,000
	Total cost (main component/BoP)					$62,500	$18,050
	Percentage (main component/BoP)					78 per cent	22 per cent
	Total cost system					$80,550	

Note

All cost numbers include installation.

Table 8.5 O&M costs for both systems (in US dollars)

		Unit	Solar cooling system	Reference system
Utilities	Electricity price (year 1)	$/kWh_{el}$	$0.2	$0.2
	Gas price (year 1)	$/kWh_{th}$	$0.05	$0.05
	Water price (year 1)	$/m^3$	$2.0	$2.0
	Annual escalation rate of electricity price	%/a	3	3
	Annual escalation rate of gas price	%/a	2	2
	Annual escalation rate of water price	%/a	1	1
Operation	Total electrical power of system (nominal)	kW_{el}	4.3	14.3
	Full-load hours of operation	h/a	2,059	2,059
	Total annual electricity consumption	kWh_{el}/a	8,853	29,373
	Total annual gas consumption	kWh_{th}/a	16,894	33,333
	Total annual water consumption	m^3/a	206	102.9
	Annual electricity cost (year 1)	$/a	$1,771	$5,875
	Annual gas cost (year 1)	$/a	$845	$1,667
	Annual water cost (year 1)	$/a	$412	$206
	Total annual operational cost (year 1)	$/a	$3,027	$7,747
Maintenance	Total annual maintenance cost (year 1)	$/a	$2,500	$2,500
O&M	Total annual O&M cost (year 1)	$/a	$5,527	$10,247

The NPV calculations in Table 8.6 are based on the following assumptions:

- year 0 is the year of installation of the system;
- operation of the system starts at year 1;
- the conventional chiller in the reference system is replaced in year 13. The capital cost for the chiller replacement has been calculated with the investment cost of year 1 ($45,000) subject to inflation rate (CPI);
- both systems operate for 20 years.

Furthermore, the NPV calculations in Table 8.6 are based on the assumptions for financial parameters and component lifetime shown in Table 8.7.

Table 8.6 Net present vaue (NPV) calculation for both systems (in US dollars)

Solar cooling system	Unit	Year 0	Year 1	Year 2	Year 3	Year 4	Year 5	Year 6	Year 7	Year 8	Year 9	Year 10	Year 11	Year 12	Year 13	Year 14	Year 15	Year 16	Year 17	Year 18	Year 19	Year 20
Capital cost	$	$170.105																				
Electricity cost	$/kWh$_{el}$		$0.20	$0.21	$0.21	$0.22	$0.23	$0.23	$0.24	$0.25	$0.25	$0.26	$0.27	$0.28	$0.29	$0.29	$0.30	$0.31	$0.32	$0.33	$0.34	$0.35
Electricity consumption	kWh$_{el}$/a		8,853	8,853	8,853	8,853	8,853	8,853	8,853	8,853	8,853	8,853	8,853	8,853	8,853	8,853	8,853	8,853	8,853	8,853	8,853	8,853
Gas cost	$/kWh$_{th}$		$0.05	$0.05	$0.05	$0.05	$0.05	$0.06	$0.06	$0.06	$0.06	$0.06	$0.06	$0.06	$0.06	$0.06	$0.07	$0.07	$0.07	$0.07	$0.07	$0.07
Gas consumption	kWh$_{th}$/a		16,894	16,894	16,894	16,894	16,894	16,894	16,894	16,894	16,894	16,894	16,894	16,894	16,894	16,894	16,894	16,894	16,894	16,894	16,894	16,894
Water cost	$/m^3		$2.00	$2.02	$2.04	$2.06	$2.08	$2.10	$2.12	$2.14	$2.17	$2.19	$2.21	$2.23	$2.25	$2.28	$2.30	$2.32	$2.35	$2.37	$2.39	$2,42
Water consumption	m^3/a		206	206	206	206	206	206	206	206	206	206	206	206	206	206	206	206	206	206	206	206
Annual maintenance cost	$/a		$2,500	$2,563	$2,627	$2,692	$2,760	$2,829	$2,899	$2,972	$3,046	$3,122	$3,200	$3,280	$3,362	$3,446	$3,532	$3,621	$3,711	$3,804	$3,899	$3,997
Annual operation cost	$/a		$3,027	$3,101	$3,177	$3,255	$3,336	$3,418	$3,503	$3,589	$3,679	$3,770	$3,864	$3,961	$4,060	$4,162	$4,266	$4,373	$4,484	$4,597	$4,713	$4,833
Annual O&M cost	$/a		$5,527	$5,664	$5,804	$5,948	$6,095	$6,247	$6,402	$6,561	$6,725	$6,892	$7,064	$7,241	$7,422	$7,608	$7,798	$7,994	$8,195	$8,401	$8,612	$8,829
Total annual cost	$/a	$170,105	$5,527	$5,664	$5,804	$5,948	$6,095	$6,247	$6,402	$6,561	$6,725	$6,892	$7,064	$7,241	$7,422	$7,608	$7,798	$7,994	$8,195	$8,401	$8,612	$8,829
Net present value (20 years)	**$**	**$262,918**																				

Reference system		Year 0	Year 1	Year 2	Year 3	Year 4	Year 5	Year 6	Year 7	Year 8	Year 9	Year 10	Year 11	Year 12	Year 13	Year 14	Year 15	Year 16	Year 17	Year 18	Year 19	Year 20
Capital cost	$	$80,550													$62,033							
Electricity cost	$/kWh$_{el}$		$0.20	$0.21	$0.21	$0.22	$0.23	$0.23	$0.24	$0.25	$0.25	$0.26	$0.27	$0.28	$0.29	$0.29	$0.30	$0.31	$0.32	$0.33	$0.34	$0.35
Electricity consumption	kWh$_{el}$/a		29,373	29,373	29,373	29,373	29,373	29,373	29,373	29,373	29,373	29,373	29,373	29,373	29,373	29,373	29,373	29,373	29,373	29,373	29,373	29,373
Gas cost	$/kWh$_{th}$		$0.05	$0.05	$0.05	$0.05	$0.05	$0.06	$0.06	$0.06	$0.06	$0.06	$0.06	$0.06	$0.06	$0.06	$0.07	$0.07	$0.07	$0.07	$0.07	$0.07
Gas consumption	kWh$_{th}$/a		33,333	33,333	33,333	33,333	33,333	33,333	33,333	33,333	33,333	33,333	33,333	33,333	33,333	33,333	33,333	33,333	33,333	33,333	33,333	33,333
Water cost	$/m^3		$2.00	$2.02	$2.04	$2.06	$2.08	$2.10	$2.12	$2.14	$2.17	$2.19	$2.21	$2.23	$2.25	$2.28	$2.30	$2.32	$2.35	$2.37	$2.39	$2,42
Water consumption	m^3/a		103	103	103	103	103	103	103	103	103	103	103	103	103	103	103	103	103	103	103	103
Annual maintenance cost	$/a		$2,500	$2,563	$2,627	$2,692	$2,760	$2,829	$2,899	$2,972	$3,046	$3,122	$3,200	$3,280	$3,362	$3,446	$3,532	$3,621	$3,711	$3,804	$3,899	$3,997
Annual operation cost	$/a		$7,747	$7,959	$8,176	$8,400	$8,630	$8,867	$9,110	$9,360	$9,617	$9,882	$10,154	$10,434	$10,721	$11,017	$11,322	$11,634	$11,956	$12,287	$12,628	$12,978
Annual O&M cost	$/a		$10,247	$10,521	$10,803	$11,092	$11,390	$11,695	$12,009	$12,332	$12,663	$13,004	$13,354	$13,714	$14,084	$14,464	$14,854	$15,255	$15,668	$16,091	$16,527	$16,974
Total annual cost	$/a	$80,550	$10,247	$10,521	$10,803	$11,092	$11,390	$11,695	$12,009	$12,332	$12,663	$13,004	$13,354	$13,714	$76,117	$14,464	$14,854	$15,255	$15,668	$16,091	$16,527	$16,974
Net present value (20 years)	**$**	**$292,916**																				

Table 8.7 Financial assumptions for NPV calculations in Table 8.6

Financial assumptions	Unit	Value
Lifetime conventional chiller[1]	a	12
Lifetime thermal chiller and solar collectors	a	20
CPI (inflation rate)	%/a	2.5%
Discount rate	%	4.0%

Note

[1] The average lifetime of the compressor for a conventional chiller is approximately eight years. Twelve years was chosen here – an optimistic assumption.

Table 8.8 Net present value comparison of solar cooling system and reference system in the above worked example (in US dollars)

Economic comparison	Unit	Value
Net present value – solar cooling system	$	$262.918
Net present value – reference system	$	$292.916
Difference in net present value = cost saved over lifetime of system	$	$29.998

Table 8.6 shows that in the worked example the solar cooling system has a lower net present value than the reference system. The NPV of each system is the total lifetime cost of each system, discounted to the year of installation. The difference in NPV is the cost saved over the lifetime of the systems – in this case 20 years – expressed as its value in the year of installation. It can be seen (see Table 8.8) that lifetime cost savings of $29,998 can be achieved by using the solar cooling system instead of the reference conventional system. It has to be stated that this is not always the case. In many projects, a conventional chiller system will be lower in cost compared to a solar cooling system.

8.1.3 Limitations

Each method of assessing economic feasibility has its limitations. The NPV method described in the previous chapter (Table 8.6) takes into account dynamic parameters such as inflation rates and rising prices for electricity and gas. It can be considered to be fairly accurate, but the results can only be as good as the inputs provided. In other words, if the input parameters have been assumed wrongly, then the NPV will represent an inaccurate value. Some key parameters that have a direct influence on the accuracy of the NPV method are:

- *Annual hours of operation*: ideally, this number is derived from annual performance simulations. If estimated manually it may be erroneous.

- *Chilled and hot water loads*: it can be difficult to obtain authoritative values, especially if the system is designed for a new building without existing building data.
- *Working life or system lifetime*: this parameter usually comes with a fluctuation margin. It influences the capital costs of the reference system.
- *Rising gas and electricity prices*: these are subject to educated guesses and can vary in reality.
- *Discount rate*: most likely the interest rate will vary over the lifetime of the system.
- *Degradation of component performance*: this has not been taken into account in this example, but will most likely happen in reality.

Despite the above limitations, the authors recommend the NPV method. It includes most of the dynamic changes (technically and financially) that take place over a system's working life, so is therefore more accurate than the simpler annuity method. In the annuity method only the first year of operation is used as a base for the economic analysis. Also, the annuity does not consider changes of boundary conditions over the system lifetime, such as operating cost increases, degradation or further capital investments.

8.2 System costs breakdown

Every solar thermal cooling system is an individual design. There are no commercially available off-the-shelf, plug-and-play solutions yet. Solar cooling kits (pre-assembled and pre-designed component groups) may come close to this, but even these kits do not result in a system which can be reproduced generally in other buildings. Buildings, and other potential applications, are far too individual and different in their requirements to make this possible. In addition, investment costs for equipment, delivery costs as well as installation costs can vary significantly depending on the project location.

The authors are therefore not in a position to present a general, detailed and accurate cost breakdown for all types of solar thermal cooling systems. However, an overview of costs for some selected representative examples can be given. Figures 8.4–8.7 present four examples, a small installation for a single-family house, a mid-size installation for a small office building, a large installation for a large office building and a very large installation for an industrial process, respectively. An overview of the costs of all four systems is given in Table 8.9.

The case studies in Chapter 10 also give cost information on different sized systems.

8.2.1 Percentage cost breakdown for some representative systems

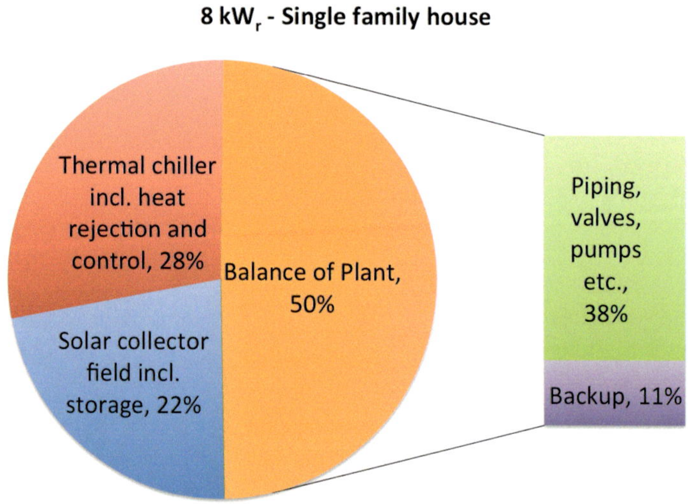

Figure 8.4 Percentage cost breakdown of a small system for a single family house. Solar field: 36 m^2, flat plate collectors. Chiller: 7.5 kW$_r$, ADsorption. Storage tank: 2.0 m^3, water.

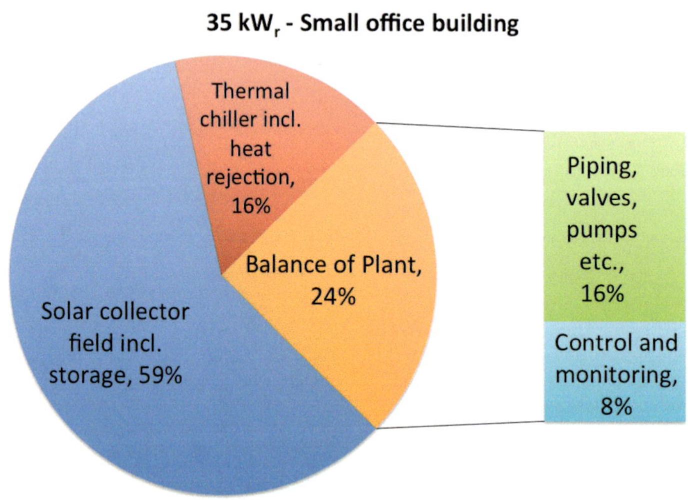

Figure 8.5 Percentage cost breakdown of a medium system for a small office building. Solar field: 90 m^2, evacuated tube collectors. Chiller: 35 kW$_r$, ABsorption. Storage tank: 1.5 m^3, water.

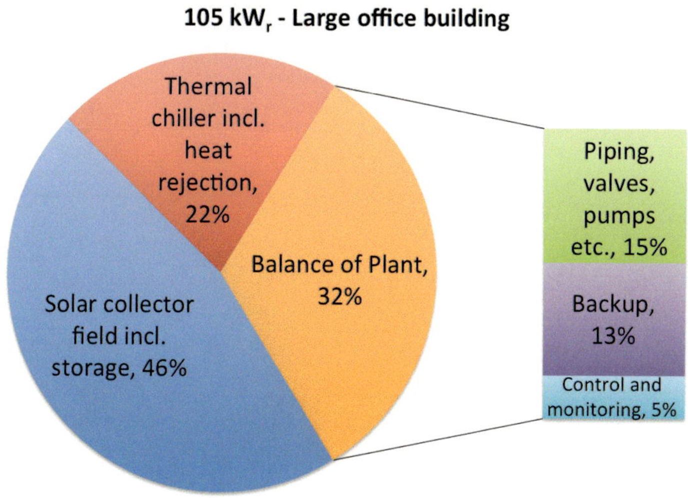

Figure 8.6 Percentage cost breakdown of a large system for a large office building. Solar field: 30 m², evacuated tube collectors. Chiller: 105 kW$_r$, ABsorption. Storage tank: none.

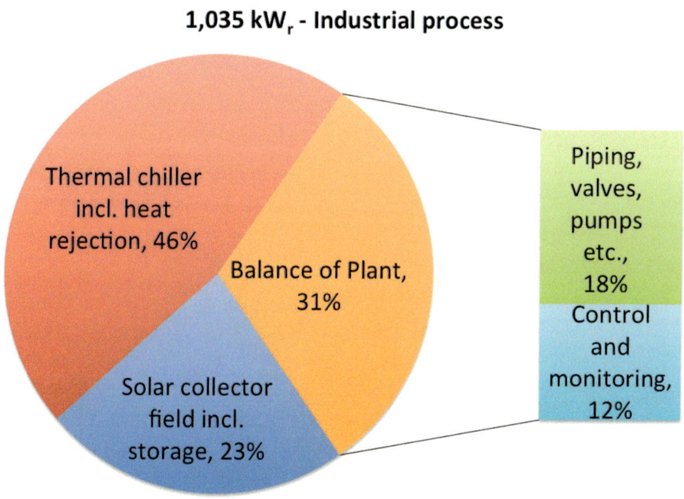

Figure 8.7 Percentage cost breakdown of a very large system for an industrial process. Solar field 1,218 m², evacuated tube collectors. Chiller: 3 · 353 kW$_r$, ADsorption. Storage tank: 17 m³, water.

Note: The solar collector field in this system is not sized to provide the required heat for all chillers. It provides a fraction of the required driving heat, the rest is provided by waste heat from an industrial process and a gas boiler.

It can be seen in Table 8.9 that the specific cost, i.e. the cost per kilowatt of cooling power installed, decreases with increasing cooling capacity.

Table 8.9 Cost breakdown for the four systems above (in Euro).

Cost breakdown	Single family house (8 kW$_r$)	Small office building (35 kW$_r$)	Large office building (105 kW$_r$)	Industrial process (1035 kW$_r$)
Balance of plant	€39,411	€68,369	€170,180	€1,001,700
Solar collector field incl. storage	€11,344	€99,582	€145,050	€301,100
Thermal chiller incl. heat rejection	€14,285	€27,240	€67,830	€601,500
Piping, valves, pumps, etc.	€19,495	€26,985	€47,400	€240,600
Back-up	€5,630	None	€39,750	None
Control and monitoring	Not installed	€14,144	€15,200	€159,600
Total cost	**€90,168**	**€236,320**	**€485,410**	**€2,304,500**
Specific cost	11,271 €/kW$_r$	6,752 €/kW$_r$	4,623 €/kW$_r$	2,195 €/kW$_r$

8.2.2 Equipment cost trends

In 2006, pre-assembled and pre-designed component groups (branded as solar cooling kits) were first introduced onto the market. This has led to a continuing decrease in cost for these kits. The use of standardised configurations and components allows for lower investment and installation costs. Figure 8.8 shows the cost development for solar cooling kits in three different capacity ranges since 2007.

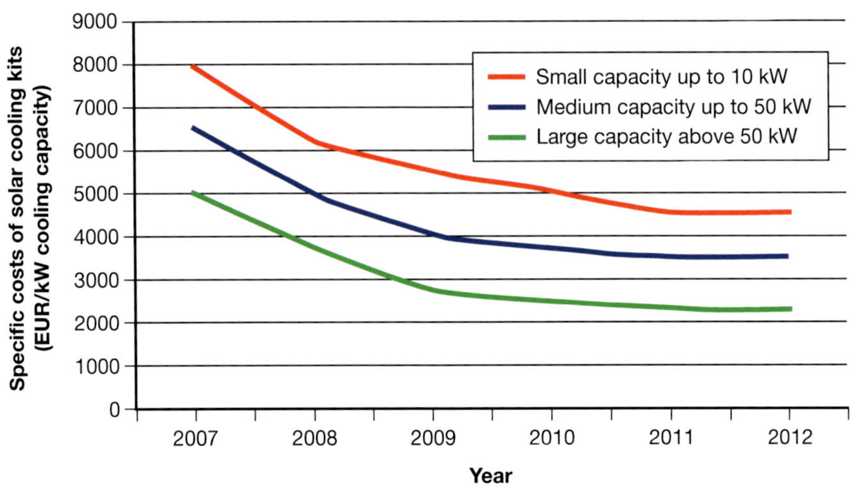

Figure 8.8 Cost development for pre-assembled and pre-designed component groups (solar cooling kits), including chiller, solar thermal collectors, pumps, valves, control, etc. Cost numbers are given ex-works and excluding VAT or other taxes.

Source: Solem Consulting

9

Potential markets for solar thermal cooling systems

There are three main potential markets in terms of building type/application for solar thermal cooling systems: industrial solar cooling systems and those for commercial buildings and private houses, and it is important to differentiate between them.

Commercial buildings (e.g. offices) typically have low heat and hot water demand but rather high air-conditioning requirements. Air-conditioning is required during weekly work hours, but not on the weekends. The geographic location as well as the building type influences economic feasibility. For example, a high-rise with a glass façade has higher air-conditioning requirements than a one-level building with concrete walls.

For *residential applications* the ratio between heat and cold requirement depends on the location. For example, a house in a temperate climate zone will have higher heat demand and lower cold demand, whereas for a house located in the tropics it will be the other way round. Operating hours of an air-conditioner vary depending on the climate zone, but are typically lower than for commercial buildings.

Industrial applications typically have both heat and cold demand with process heating or chilling being the main requirement. Air-conditioning may only be a small part of a factory's overall heat and cold requirement, with most of this heat/cold requirement determined by machinery or process operation. Machinery and processes typically operate all year round, and therefore the absolute heat/cold demand is higher than for commercial buildings or private houses. This influences the economics positively. The geographic location has less influence on economic feasibility.

Note: PV-cooling systems are not discussed in this chapter. Please refer to Chapter 3.5 for details on these.

9.1 Commercial buildings and residential houses

The applicability of solar thermal cooling in commercial and private buildings depends on the local geographic and economic conditions. There are three main factors influencing the economic feasibility of solar thermal cooling systems at any given location:

- *solar resource*: the available solar radiation determines the potential system performance;
- *climate zone*: seasonal and daily variations of local climate determine the heating and cooling requirements/loads;
- *energy costs*: electricity and fossil fuel costs determine the economic feasibility.

A fourth, less important, factor is the daily and annual simultaneity of solar radiation levels with cooling requirements. In other words, a building's heating/cooling loads should coincide with the times when the sun is shining (daytime/middle of the day), and not be early in the morning, late in the afternoon or at night. The simultaneous presence of solar energy and cooling requirement reduces the need for storage volume, which of course lowers system costs. These factors are discussed in more detail below.

9.1.1 Solar resource

A solar cooling system requires good solar radiation conditions. Typically, the annual global insolation on the horizontal is used for initially assessing this parameter. Solar irradiation levels greater than 1,600 kWh/(m^2 a) provide good conditions for the operation of a solar cooling system. Figure 9.1 shows the world map with satellite-based data of the global annual irradiance.

However, not all locations on Earth with global solar insolation greater than 1,600 kWh/(m^2 a) are equally suitable for solar cooling. Take the tropics as an example. Figure 9.1 shows that the tropical regions receive approximately 1,700–2,000 kWh/(m^2 a) of global insolation. However, most of this is diffuse radiation due to the high cloud cover in these areas and thus not suitable for solar thermal collectors that require direct radiation.

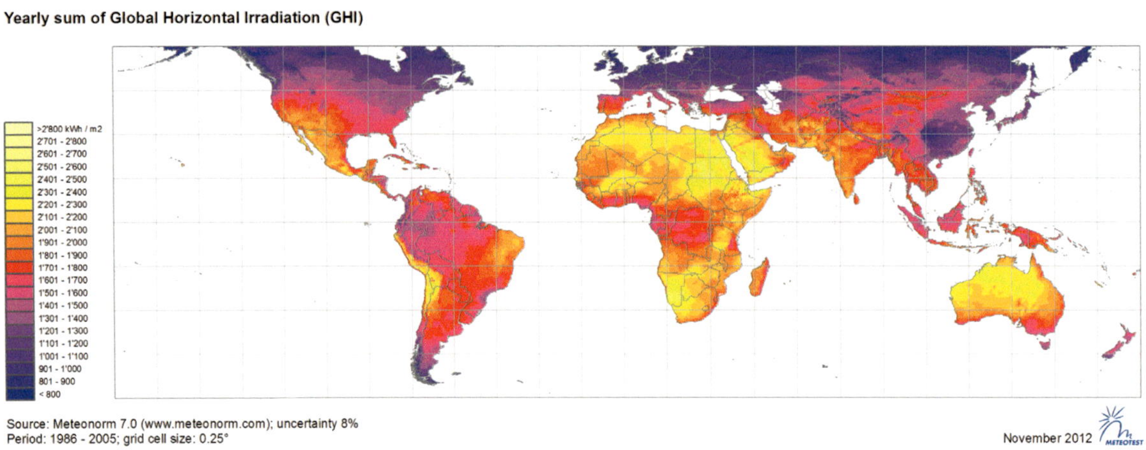

Figure 9.1 Global horizontal irradiation (GHI) distribution.

Source: © METEOTEST, www.meteonorm.com

Figure 9.2 Global average cloud cover in June 2013. Dark blue areas are virtually cloud-free; white areas have very high cloud coverage.

Source: NASA

Figure 9.3 Global average cloud cover in December 2012. Dark blue areas are virtually cloud-free; white areas have very high cloud coverage.

Source: NASA

Figures 9.2 and 9.3 show the global monthly average cloud coverage for June 2013 and December 2012, respectively. It can be seen that the tropics have a mean annual cloud cover of approximately 0.8–1.0 (80–100 per cent). There is a dense band of clouds around the Equator all year round (also known as the intertropical convergence zone). Dense cloud cover means less solar radiation reaching the ground and a high percentage of diffuse radiation. Both of these effects will result in lower performance of all solar thermal collectors. Also, the tropics typically receive a midday or afternoon shower during monsoon season, reducing solar irradiation and increasing humidity. The performance of a solar cooling system under these conditions (low solar resource, high ambient humidity) will be very limited.[1] It is therefore essential to accurately assess the solar potential carefully for each location.

If concentrating collectors are to be used in a solar cooling system then this becomes even more essential. Concentrating collectors can only use direct radiation – they don't work at all with diffuse radiation, as it cannot be reflected from the mirrors. Figure 9.4 shows the global distribution of direct normal irradiation (DNI). It can be seen that operation of a solar cooling system using concentrating collectors is not really possible in the tropics, as annual DNI is too low. Locations with good DNI potential are those with DNI greater than 1,700 kWh/(m^2 a).

9.1.2 Climate zone

The seasonal variations within a climate zone have a strong influence on the performance of a solar cooling system. Climate zones with non-incisive summer–winter differences, as in the subtropical or tropical zones, typically

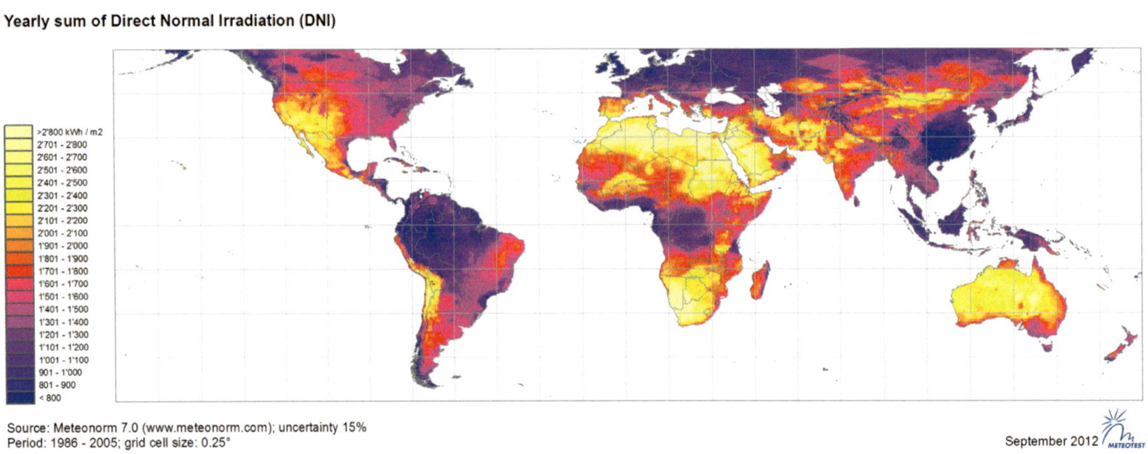

Figure 9.4 Distribution of global direct normal irradiation (DNI) on the horizontal plane.

Source: © METEOTEST; based on www.meteonorm.com

1 High ambient humidity results in high wet bulb temperatures, limiting the cooling performance of wet cooling towers due to limited water evaporation.

incur a high annual cooling load and a low annual heating load. Temperate zones with a stronger summer–winter variation incur a lower cooling load than the tropics but a much higher heating load. A solar cooling system will provide the best economic performance if the solar array (one of the highest cost factors) can be utilised as much as possible, so from the point of view of economic feasibility it is imperative to either have:

- high annual cooling loads (system provides solar cooling only); or
- combined annual heating and cooling loads (system provides solar heating and cooling).

Preferred climate zones for solar thermal cooling are therefore those with very high cooling loads (the Equatorial, tropic and subtropic zones) and those with high combined summer cooling and winter heating loads (the temperate zones). Figures 9.5 and 9.6 show the global distribution of average annual cooling and heating degree days. These two parameters provide a measure of air-conditioning and space-heating demand of private houses, respectively, on an annual basis. Countries with high numbers of cooling days (a high annual cooling load) typically have low numbers of heating days (a low annual heating load).

Regions where the combination of heating and cooling load looks promising for the economic operation of solar cooling systems include North America, the south of Europe, the north and south of Africa, Australia, the north and south of South America, parts of the Caribbean and China.

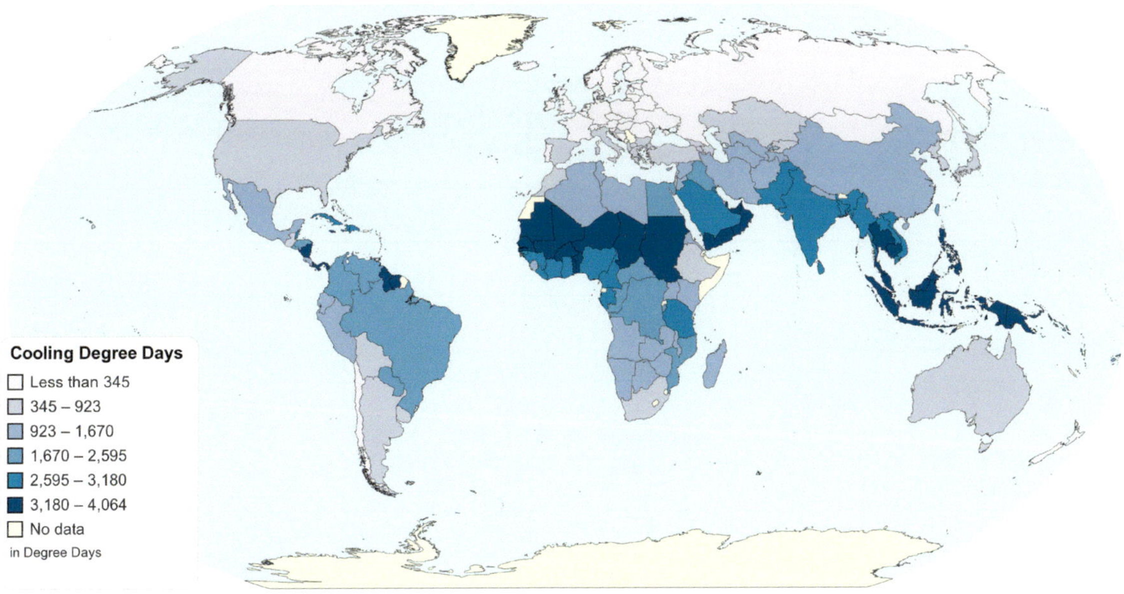

Figure 9.5 Global distribution of average annual cooling degree days (CDDs).

Source: Chartsbin

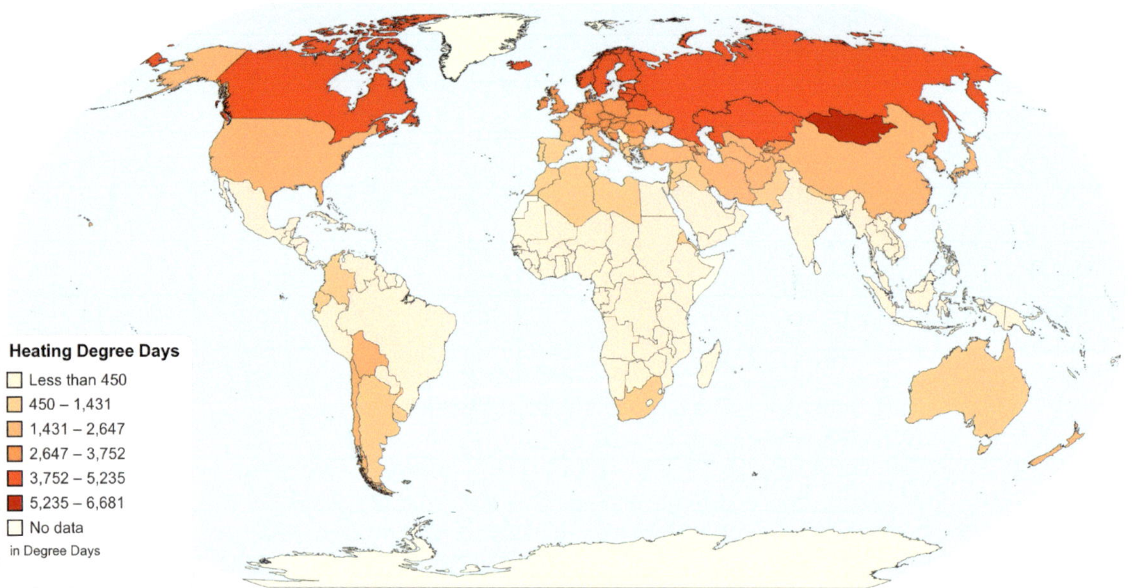

Figure 9.6 Global distribution of average annual heating degree days (HDDs).

Source: Chartsbin

9.1.3 Energy costs

The local cost of electricity, natural gas, oil, water[2] and even diesel[3] determine the operational cost of any heating or cooling system. This applies to both standard and solar-driven systems. As discussed in Chapter 8, the economic feasibility of solar cooling depends on the savings made compared to a standard reference system. Therefore, the higher the cost of utilities at a given location, the higher the savings that can be made by a solar system. The main cost drivers are electricity and natural gas, since these are typically used to power air-conditioners and gas heaters. Oil and diesel may play a role in some applications but shall not be discussed in detail here. Figure 9.7 shows a global overview of residential electricity prices. Countries with relatively high electricity prices are more favourable for solar cooling. These include: Central and southern Europe, Mexico, Central America, Brazil, Australia and some parts of Asia.

Natural gas as a piped resource is not available in most countries. A collection of global retail gas prices in the residential sector is given in Figure 9.8.

[2] Water is required for the operation of a wet cooling tower.

[3] Diesel generators can be used to generate electricity for air-conditioning, e.g. for stand-alone systems.

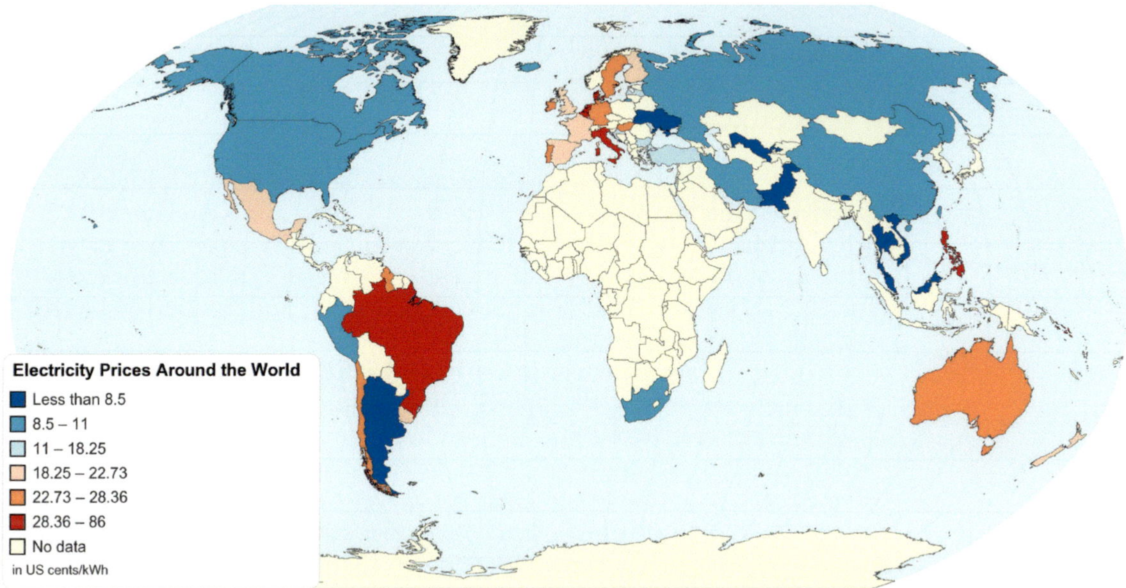

Figure 9.7 Global distribution of 2012 residential electricity prices in US$/kWh, excl. VAT/local taxes.

Source: Chartsbin

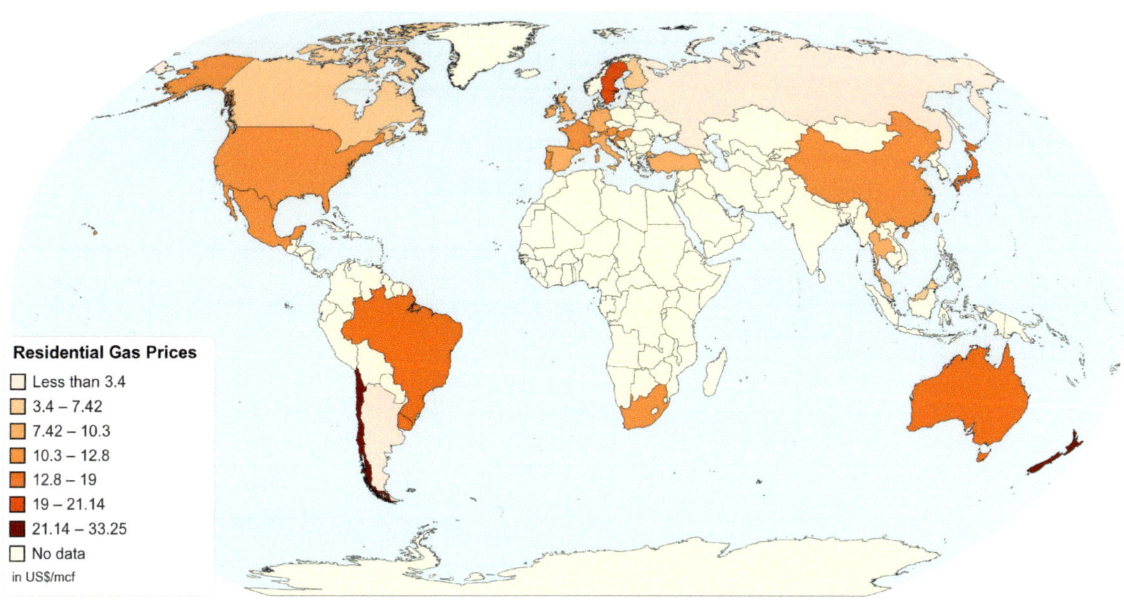

Figure 9.8 Global distribution of 2011 residential retail gas prices in US$, excl. VAT/local taxes.

Source: Chartsbin

Countries with rather high gas prices include Brazil, Chile, Australia, New Zealand and Japan, among others.

9.1.4 Summary

Using the information given above the main markets for residential and commercial thermal solar cooling can thus be deducted. Combining the three main drivers affecting the economic feasibility of a solar cooling system, a matrix showing the influence of each parameter can be constructed (see Table 9.1). However, this list is not comprehensive, and data are not currently available for all countries.

In order to rate the countries in Table 9.1 for their economic feasibility, a score point ranking can be applied. The lowest numerical value per driver in Table 9.1 has been accounted for with five points,[4] the highest value with ten points. Intermediate values have been interpolated accordingly (Table 9.2). For example: the solar resource in Table 9.1 ranges from 1,500 kWh/(m^2 a) (Japan) to 2,300 kWh/(m^2 a) (Chile). Therefore Chile achieves ten points for its solar resource and Japan five points. Score points for all other countries with values in between have been linearly interpolated. This procedure has been applied to all other drivers (Table 9.2).

Table 9.1 Values of main economic drivers for selected countries

	Solar resource (average) (kWh/(m^2 a))	**Heating degree days (d/a)**	**Cooling degree days (d/a)**	**Electricity cost (USct/ kWh)**	**Natural gas cost (US$/mcf)**
Australia	2,200	828	839	25.0	19.1
Brazil	2,000	118	2,015	34.2	18.1
Chile	2,300	1,613	225	23.1	33.2
China	1,800	2,158	1,046	9.0	11.8
France	1,600	2,478	241	19.4	10.9
Italy	1,700	1,838	600	28.4	10.0
Japan	1,500	1,901	896	30.4	17.2
North America	1,800	2,159	882	11.0	12.1
South Africa	2,000	630	824	11.0	12.3
Spain	1,700	1,431	702	22.7	9.1
Uruguay	2,000	1,019	732	19.0	18.2

4 A value of 5 points has been chosen instead of zero to account for the fact that even countries with low absolute driver values may allow for economic operation.

Table 9.2 Score ratings based on main economic drivers for selected countries

	Solar resource (score)	Sum of heating and cooling degree days (score)	Electricity cost (score)	Natural gas cost (score)	Total Score
Australia	9.4	5.4	8.2	7.1	30.0
Brazil	8.1	6.7	10.0	6.9	31.7
Chile	10.0	5.9	7.8	10.0	33.6
China	6.9	9.8	5.0	5.6	27.2
France	5.6	8.4	7.1	5.4	26.4
Italy	6.3	7.6	8.8	5.2	27.8
Japan	5.0	8.6	9.2	6.7	29.5
North America	6.9	9.3	5.4	5.6	27.2
South Africa	8.1	4.8	5.4	5.7	23.9
Spain	6.3	6.7	7.7	5.0	25.7
Uruguay	8.1	5.6	7.0	6.9	27.6

It can be seen from Table 9.2 that the top four countries favourable for the economic operation of a solar thermal cooling system are Chile, Brazil, Australia and Japan. There, the combination of good solar resource, large thermal loads and rather high energy costs provides the most attractive economic conditions. The authors, however, advise that this result is to be taken with some caution. The data in Table 9.1 have been collected to the best of our knowledge, but it has been averaged per country and thus certainly has an error range.

In summary, the conditions favouring the economic operation of a solar cooling system are:

- high solar resource (global horizontal irradiation >1,600 kWh/(m^2 a), DNI >1,700 kWh/(m^2 a) if concentrating collectors are to be used);
- low annual average cloud cover (<50 per cent);
- high annual thermal load of building, comprising both heating/hot water and cooling loads;
- high energy costs (mainly electricity and natural gas);
- simultaneity of solar radiation and building thermal loads to minimise storage volumes.

Furthermore, the project location should have sufficient infrastructure to enable the delivery, construction, installation and commissioning of the system. Qualified technicians should be available for the installation, maintenance and servicing. Experience with existing systems has shown that the latter is not always the case. The installation of a solar thermal cooling system usually requires a greater skill set than a standard solar hot water system. Therefore it is recommended to choose a complete package provider, companies that provide turn-key solutions, ranging from planning and design to installation and commissioning. The fewer companies involved in the process, the better. See Chapter 7 for recommendations regarding system installation.

9.2 Industrial applications

Typical industrial applications for solar thermal cooling are those with high combined heat and cold loads, providing high use (a high solar fraction) of the solar resource available. A non-comprehensive list of industrial applications that have been identified as promising for solar heating and cooling is given below.

Promising industrial applications for solar thermal cooling

- bakeries
- meat-processing plants
- dairy industry
- wineries
- fish and seafood processing plants
- biogas plants
- electroplating industry

A complete analysis of all industry sectors for the applicability of solar cooling would exceed the scope of this book. A market analysis of this sector for Europe from 2008 is available for download (see [19]). Also, the case studies in Chapter 10 of this book provide some information on industrial solar thermal cooling systems.

9.3 Overview of commercially available thermal chillers

Tables 9.3–9.5 give an overview of commercially available small- (<20 kW_r) and medium-scale (20–200 kW_r) thermal chillers. It is a snapshot in time, with prices current at the time of writing. The list is not exhaustive. All prices are given ex-works in euros, US dollars and Australian dollars, excl. VAT/other taxes (where applicable), transport and installation cost.

Updates to this chiller list will be provided at regular time intervals. Please contact Solem Consulting if you're interested in receiving an updated version: info@solem-consulting.com.

Table 9.3 Overview of commercially available small-scale thermal sorption chillers (<20 kW$_r$)

	Manufacturer and country of origin																
	EAW (DE)	**Fischer Eco Solutions (DE)**	**InvenSor (DE)**	**InvenSor (DE)**	**InvenSor (DE)**	**Jiangsu Huineng (CN)**	**Jiangsu Huineng (CN)**	**Mitsubishi Plastics (JP)**	**Mitsubishi Plastics (JP)**	**Pink (AT)**	**Sakura (JP)**	**Sakura (JP)**	**SolabCool (NL)**	**SorTech (DE)**	**SorTech (DE)**	**Thermax (IND)**	**Yazaki Energy (J)**
Product name	Wegracal SE 15, chillii® ESC15	N.A.	chillii® ISC10, LTC10	LTC10 e	chillii® ISC18, HTC18	RXZ-11	TX-11	U-type	M-type	chillii® PSC 19, Pink Chiller PC19	SHL003	SHL003	SolabChiller	chillii® STC8, ASC08	chillii® STC15, ASC15	LT 0.5	WFC-SC5, chillii® WFC18
Distributors	EAW, SolarNext Solution	Fischer Eco Solutions	SolarNext (R), InvenSor (WS)	InvenSor (WS)	SolarNext (R), InvenSor (WS)	Huineng	Huineng	Mitsubishi Plastics	Mitsubishi Plastics	SolarNext (R), Pink (WS)	Sakura	Sakura	SolabCool	SolarNext (R), SorTech (WS)	SolarNext (R), SorTech (WS)	Thermax, Air Solutions	Yazaki Europe, SolarNext, Sonnen-kraft, York
Sorbent/ refrigerant	$LiBr/H_2O$	$LiBr/H_2O$	Zeolithe/ H_2O	Zeolithe/ H_2O	Zeolithe/ H_2O	$LiBr/H_2O$	$LiBr/H_2O$	Zeolithe/ H_2O	Zeolithe/ H_2O	H_2O/NH_3	$LiBr/H_2O$	$LiBr/H_2O$	$LiCl/H_2O$	Silicagel/ H_2O	Silicagel/ H_2O	$LiBr/H_2O$	$LiBr/H_2O$
Nominal cooling capacity	15 kW	15 kW	10 kW	9 kW	18 kW	11 kW	11.5 kW	10 kW	10 kW	19 kW	10.5 kW	17.5 kW	5 kW	8 kW	15 kW	17.5 kW	17.5 kW
Nominal hot water heat input	21 kW	21 kW	16.7 kW	12.9 kW	34.6 kW	15.7 kW	16.4 kW	20 kW	21.8 kW	30 kW	14.6 kW	24.5 kW	8.3 kW	14 kW	27 kW	25 kW	25 kW
Nominal hot water temperature	90 °C	85 °C	72 °C	72 °C	85 °C	90 °C	90 °C	70 °C	70 °C	85 °C	88 °C	88 °C	55 °C	75 °C	75 °C	85 °C	88 °C
Nominal chilled water temperature	10–15 °C	10–15 °C	15 °C	15 °C	15 °C	10 °C	10 °C	15 °C	11 °C	6 °C	8 °C	8 °C	15 °C	15 °C	15 °C	6-8 °C	6-8 °C
Nominal cooling water temperature	27 °C	27 °C	27 °C	27 °C	27 °C	30 °C	–	–	32 °C	24 °C	31 °C	31 °C	–	27 °C	27 °C	30 °C	31 °C
Size	1.75 · 0.8 · 1.75 m	N.A.	1.1 · 1.37 · 0.75 m	1.1 · 1.37 · 0.75 m	1.1 · 1.37 · 0.75 m	0.6 · 0.7 · 1.9 m	1.67 · 1.67 · 2.13 m	2.26 · 0.95 · 2.4 m	1.34 · 0.8 · 1.15 m	0.8 · 0.6 · 1.9 m	1.28 · 1.26 · 1.58 m	1.28 · 1.26 · 1.58 m	1.00 · 0.65 · 1.20 m	0.79 · 1.06 · 0.94 m	0.79 · 1.34 · 1.39 m	2.0 · 1.0 · 0.9 m	1.82 · 0.74 · 0.6 m
Weight	700 kg	N.A.	390 kg	440 kg	420 kg	400 kg	1000 kg	1000 kg	555 kg	550 kg	400 kg	500 kg	–	265 kg	530 kg	1300 kg	420 kg
Unit cost US$[+]	32,529	N.A.	33,768[#]	N.A.	52,021[#]	N.A.	N.A.	N.A.	N.A.	43,025[#]	N.A.	N.A.	N.A.	28,553[#]	49,413[#]	61,811	24,641[#]
Unit cost EUR[+]	24,950	N.A.	25,900[#]	N.A.	39,900[#]	N.A.	N.A.	N.A.	N.A.	33,000[#]	N.A.	N.A.	N.A.	21,900[#]	37,900[#]	47,409	18,900[#]
Unit cost AU$[+]	31,622	N.A.	32,826[#]	N.A.	50,570[#]	N.A.	N.A.	N.A.	N.A.	41,825[#]	N.A.	N.A.	N.A.	27,757[#]	48,035[#]	60,087	23,954[#]

Notes

All prices are ex-works and exclude GST/VAT, transport and installation cost except where otherwise stated.
Unit cost includes chiller only except where stated otherwise.
+ Exchange rates calculated as per 17 April 2013.
# Price includes chiller, cooling tower, system controller, two pumps, valves (complete kits available from retailer SolarNext only).
Abbreviations: R: retailer; WS: wholesaler

Table 9.4 Part I: overview of commercially available medium-scale thermal sorption chillers (20–200 kW_r)

	Manufacturer and country of origin											
	AGO (GER)**	BROAD (PRC)	EAW (GER)	EAW (GER)	Energy Concepts LLC (USA)	Energy Concepts LLC (USA)	HIJC (USA)	HIJC (USA)	Mayekawa (J)	Mayekawa (J)	Solarice (GER)***	Solarice (GER)****
Product name	chillii® ACC50, congelo 50	BH20	Wegracal SE 30, chillii® ESC30	Wegracal SE 50, chillii® ESC50	TS 25	TS50	NADAC-020	NADAC-050	Z-3515	Z-3525	AAC25	AAC40
Distributors	SolarNext, AGO	BROAD, Energy Conservation Systems	EAW, SolarNext, Solution	EAW, SolarNext, Solution	Energy Concepts	Energy Concepts	GBU (GER), Icogen (E)	GBU (GER), Icogen (E)	Mayekawa, Mycom S.A. (EU)	Mayekawa, Mycom S.A. (EU)	Solarice	Solarice
Sorbent/refrigerant	Water/NH_3	LiBr/H_2O	LiBr/H_2O	LiBr/H_2O	Water/NH_3	Water/NH_3	Silicagel/H_2O	Silicagel/H_2O	Silicagel/H_2O	Silicagel/H_2O	Water/NH_3	Water/NH_3
Nominal cooling capacity	50 kW	233 kW	30 kW	50 kW	87 kW	175 kW	69 kW	200 kW	105 kW	215 kW	25 kW	40 kW
Nominal hot water heat input	92 kW	188 kW	40 kW	72 kW	145 kW	277 kW	142 kW	340 kW	175 kW	360 kW	50 kW	53 kW
Nominal hot water temperature	95 to 115 °C	180 °C	90 °C	86 °C	Steam at 155 °C/5.5 bar	Steam at 155 °C/5.5 bar	90 °C	90 °C	68 °C	68 °C	95 °C	95 °C
Nominal chilled water temperature	−10 to + 6 °C	7 °C	11 °C	9 °C	12 °C	12 °C	7 °C	7 °C	15 °C	15 °C	−3 °C	10 °C
Nominal cooling water temperature	25 °C	30 °C	30 °C	27 °C	20 °C##	20 °C##	29 °C	29 °C	32 °C	32 °C	24 °C	20 °C
Size (L · W · H)	2.35 · 2.63 · 1.35 m	2.8 · 2.2 · 1.6 m	2.2 · 0.79 · 2.14 m	2.31 · 1.10 · 2.95 m	0.9 · 0.9 m footprint	1.2 · 1.2 m footprint	2.7 · 2.3 · 1.5 m	3.4 · 2.6 · 2.0 m	2.2 · 1.93 · 3.1 m	2.4 · 1.6 · 2.2 m	2.0 · 1.4 · 1.4 m	2.0 · 1.4 · 1.4 m
Weight	1600 kg	5000 kg	1400 kg	2250 kg	544 kg	1100 kg	5000 kg	7500 kg	5500 kg	5500 kg	950 kg	950 kg
Unit cost US$+	116,037	142,987	48,794	79,042	60,365	81,747	N.A.	N.A.	218,081	298,318	104,302	104,302
Unit cost EUR+	89,000	109,671	37,425	60,625	46,300	62,700	N.A.	N.A.	167,268	228,810	80,000	80,000
Unit cost AU$+	112,801	139,000	47,433	76,838	58,682	79,468	N.A.	N.A.	212,000	290,000	101,394	101,394

Notes

All prices are ex-works and exclude GST/VAT, transport and installation cost except where otherwise stated.
Unit cost includes chiller only except where stated otherwise.
+ Exchange rates calculated as per 17 April 2013.

Input temperatures are given for 10–15 °C chilled water temperature and 27–30 °C cooling water temperature (wet heat rejection) unless stated otherwise:
* Input temperatures given for 7 °C chilled water temperature and 31 °C cooling water temperature.
** Input temperatures given for −10 °C chilled water temperature and 25 °C cooling water temperature.
*** Input temperatures are given for −3 °C chilled water and 24 °C cooling water temperature.
**** Input temperatures are given for 12 °C chilled water and 24 °C cooling water temperature.
# Price includes chiller, cooling tower, system controller, two pumps, valves (complete kits available from retailer SolarNext only).
Cooling water inlet temperature 20 °C (mains water) for hot water production, chilled water 6–12 °C.
Discontinued product at time of printing, unit cost includes transport and 12 month warranty/service and commissioning.

Table 9.5 Part II: overview of commercially available medium-scale thermal sorption chillers (20–200 kW_r)

	Manufacturer and country of origin							
	Thermax (IND)*	**Thermax (IND)***	**Thermax (IND)***	**Thermax (IND)***	**Yazaki Energy (J)***	**Yazaki Energy (J)***	**Yazaki Energy (J)***	**Yazaki Energy (J)***
Product name	LT 1C ###	LT 2C	LT 3C	LT 5C	WFC-SC 10 chillii® WFC35	WFC-SC 20 chillii® WFC70	WFC-SC 30 chillii® WFC105	WFC-SC 50 chillii® WFC175
Distributors	Thermax, Air Solutions (AUS)	Thermax, Air Solutions (AUS)	Thermax, Air Solutions (AUS)	Thermax, Air Solutions (AUS)	Yazaki Europe, York, SolarNext	Yazaki Europe, York, SolarNext	Yazaki Europe, York, SolarNext	Yazaki Europe, York, SolarNext
Sorbent/refrigerant	$LiBr/H_2O$	$LiBr/H_2O$	$LiBr/H_2O$	$LiBr/H_2O$	$LiBr/H_2O$	$LiBr/H_2O$	$LiBr/H_2O$	$LiBr/H_2O$
Nominal cooling capacity	35 kW	70 kW	105 kW	175 kW	35 kW	70 kW	105 kW	175 kW
Nominal hot water heat input	50 kW	102 kW	150 kW	250 kW	50 kW	100 kW	151 kW	251
Nominal hot water temperature	70–110 °C	90 °C	90 °C	90 °C	88 °C	88 °C	88 °C	88 °C
Nominal chilled water temperature	7 °C	7 °C	7 °C	7 °C	7 °C	7 °C	7 °C	7 °C
Nominal cooling water temperature	29 °C	29 °C	29 °C	29 °C	31 °C	31 °C	31 °C	31 °C
Size (L · W · H)	2.1 · 1.6 · 1.6 m	2.2 · 1.8 · 1.4 m	2.3 · 2.1 · 1.4 m	2.4 · 2.5 · 1.5 m	2.0 · 0.9 · 1.0 m	2.1 · 1.4 · 1.2 m	2.1 · 1.5 · 1.5 m	1.8 · 2.0 · 2.1 m
Weight	2000 kg	2500 kg	2990 kg	3990 kg	600 kg	1155 kg	1800 kg	2100 kg
Unit cost US$+	78,180	88,467	94,639	102,354	39,896	53,977	64,537	169,492#
Unit cost EUR+	59,964	67,854	72,588	78,506	30,600	41,400	49,500	130,000#
Unit cost AU$+	76,000	86,000	92,000	99,500	38,783	52,471	62,738	164,766#

Notes

All prices are ex-works and exclude GST/VAT, transport and installation cost.
Unit cost includes chiller only except where stated otherwise.
+ Exchange rates calculated as per 17 April 2013
Input temperatures are given for 10–15 °C chilled water temperature and 27–30 °C cooling water temperature (wet heat rejection) unless stated otherwise:
* Input temperatures given for 7 °C chilled water temperature and 31 °C cooling water temperature.
** Input temperatures given for –10 °C chilled water temperature and 25 °C cooling water temperature.
*** Input temperatures are given for –3 °C chilled water and 24 °C cooling water temperature.
**** Input temperatures are given for 12 °C chilled water and 24 °C cooling water temperature.
# Price includes chiller, cooling tower, system controller, two pumps, valves (complete kits available from retailer SolarNext only).
Cooling water inlet temperature 20 °C (mains water) for hot water production, chilled water 6–12 °C.
Discontinued product at time of printing, unit cost includes transport and 12 month warranty/service and commissioning.

10
Case studies

In this chapter several existing solar thermal cooling installations are presented as case studies. The technical design parameters are given, as well as hydraulic layouts. Experience with the system is described in anecdotal form.

10.1 Steinway & Son piano factory, New York City, USA

This system consists of 38 concentrating parabolic trough collectors with a double-effect ABsorption chiller, which controls the internal office building climate and provides steam for the Steinway & Son Piano manufacturing facility in Long Island, New York City, USA [20, 21]. The project is a collaboration between Steinway & Son, ERS, Sustainable Energy Consulting, Abengoa Solar Inc., Sunshine Plus Solar, BROAD USA and Schuyler Engineering. The solar system produces process heat and cooling and has been operational since 2010.

Figure 10.1 Rooftop installation of parabolic trough collectors at the New York Steinway & Son factory.

Source: Henkel Solar, Inc

Figure 10.2 Double-effect ABsorption chiller installed at the Steinway & Son piano factory.

Source: Henkel Solar, Inc

Table 10.1 shows the specifications of the components of the system. The system layout is shown in Figure 10.3.

Table 10.1 Component specifications for New York Steinway & Son piano factory solar system

Solar collectors	**Parabolic trough collector**
Model	Abengoa IST PT-1
Total aperture area	532 m^2
Max. working temperature	200 °C (392 °F)
Collector fluid	Pressurised water
Chiller	**ABsorption**
Type	Water/LiBr, double-effect
Model	BROAD BZH (dual fuel source, hot water and natural gas)
Nominal chilling power output	346 kW_r (99 RT)
Nominal hot water power input	266 kW_{th}
Nominal COP	1.3
Hot water input temperature	180 °C (356 °F)
Chilled water output temperature	7 °C (44.6 °F)
Heat rejection	
Type	Wet cooling tower
Capacity	639 kW_{th}
Back-up auxiliary energy source	
Type	AERCO steam boiler

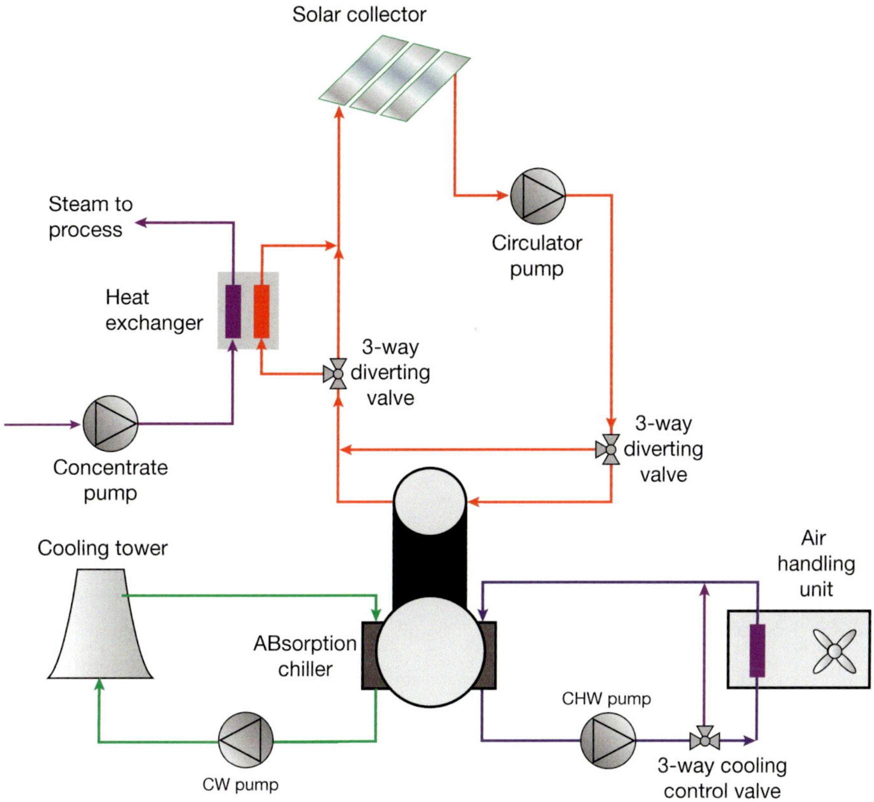

Figure 10.3 Basic system configuration of Steinway & Son piano factory system.

The solar system consists of 532 m^2 parabolic trough collectors which provide direct heat via a heat exchanger for a double-effect water/lithium bromide ABsorption chiller. The chilled water from the ABsorption chiller is distributed to an air-handling unit which then cools the air supply to the building to provide a consistent level of humidity and temperature in the factory halls.

The estimated annual performance of the solar collectors is given in Figure 10.4 and shows the proportion of useful solar thermal production the collectors have contributed to either process heat or cooling the building. The annual useful gain predicted from collectors is 430 MWh$_{th}$ [22]. The annual energy saving predicted per year is 376 MWh$_{th}$. The capacity of the ABsorption chiller over seven days in July is shown in Figure 10.5. The maximum air-conditioning power achieved during the day is 280 kW$_r$ (80 RT).

The costs of the system were approximately €644,350, including upgrading of the air-distribution network within the building [20]. The ABsorption chiller and collector costs were approximately 50 per cent of the total. The breakdown of funding sources and annual savings for the project is given below (this information is from pre-installation analysis; post-installation is not available).

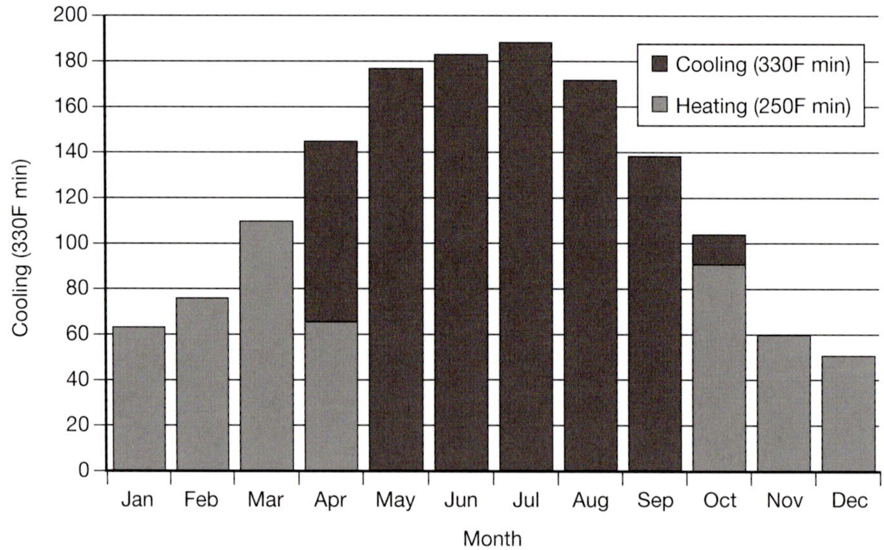

Figure 10.4 Total useful solar thermal production of the Steinway & Son piano factory system.

Source: Henkel Solar, Inc

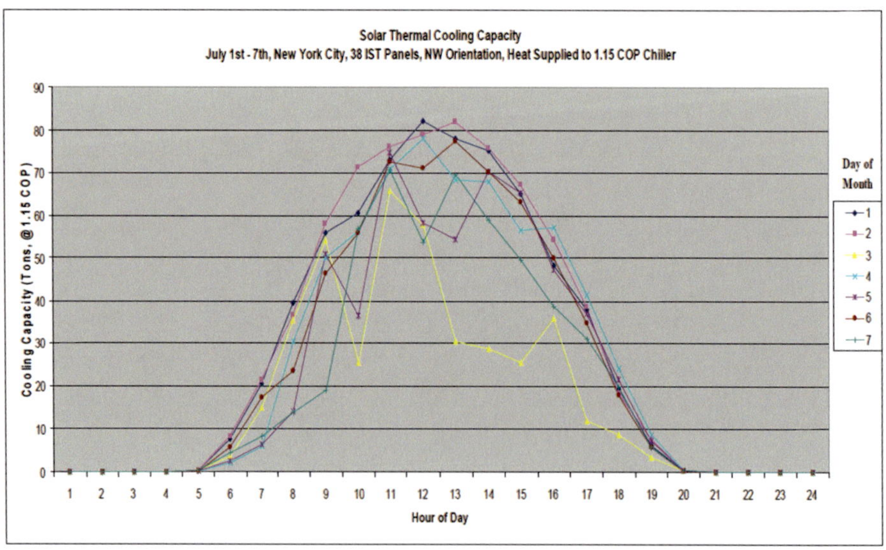

Figure 10.5 Predicted cooling capacity of a seven-day period, 1–7 July, of the Steinway & Son system.

Source: Henkel Solar, Inc

Incentives provided [22]:

- New York State Energy Research and Development Authority (NYSERDA) grant (30 per cent): €220,920;
- Federal five-year Modified Accelerated Cost-Recovery Systems (MACRS) with 50 per cent first-year bonus: €195,880;
- various New York State (NYS) small tax incentives;
- first-year sale of Solar Renewable Energy Certificates (SRECs) at €18.41/ MWh_{th}: €6,935.

The annual financial energy savings and maintenance costs for the system are €36,820 and €2,915, respectively. The specific costs for the installed solar system were €1,862/kW_r cooling capacity and €1,211/m^2 collector surface.

10.2 Grombalia winery, Tunisia

This system, at a winery in Grombalia, Tunisia, north Africa, cools stored wine using four modules of Fresnel collectors and an ammonia/water ABsorption chiller. The climate is hot desert. The country's sunshine rate exceeds 3,000 sun hours per year. The project was a collaboration between POLIMI, ANME, Domain Nefris, PSE/Mirroxx and Electrosystem, and has been operational since 2008 [23].

Table 10.2 shows system component specifications. The system layout is presented in Figure 10.8.

Figure 10.6 Wine storage tanks at Grombalia, Tunisia.

Source: POLIMI

Figure 10.7 The Grombalia solar cooling system consists of a small-scale, directly air-cooled ammonia/water ABsorption chiller with Fresnel collectors as heat source.

Source: POLIMI

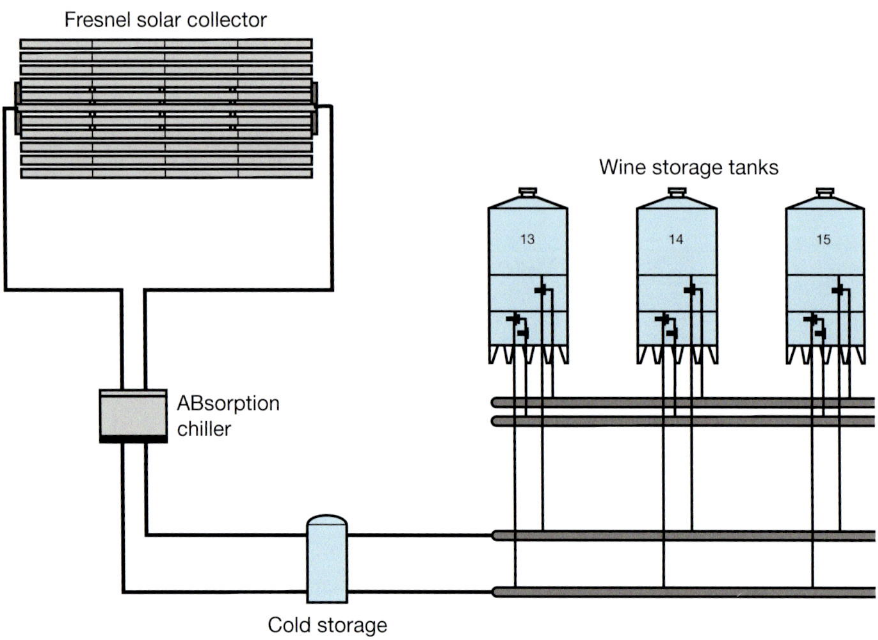

Figure 10.8 Basic system layout of the directly air-cooled ammonia/water ABsorption system at the Grombalia winery.

Source: POLIMI

Table 10.2 Component specifications for winery solar cooling system in Tunisia

Solar collectors	Fresnel collectors
Model	Mirroxx (now Industrial Solar)
Total aperture area	88 m^2
Max. pressure	16 bar
Max. temperature	200 °C (392 °F)
Max. working temperature	180 °C (356 °F)
Collector fluid	Pressurised water
Chiller	**ABsorption**
Type	Ammonia/water, single-effect
Model	Robur ACF 60-00 LB
Nominal chilling power output	12.8 kW$_r$
Nominal hot water power input	23.8 kW$_{th}$
Nominal COP	0.54
Hot water input temperature	180 °C (356 °F)
Chilled water output temperature	–10 °C (14 °F)
Heat rejection	
Type	Directly air-cooled chiller
Capacity	36.6 kW$_{th}$
Cold storage	
Capacity	3 m^3
Fluid	Water–glycol
Back-up auxiliary energy/cooling source	
Type	Vapour compression chiller (electrically powered)

The solar system consists of 88 m^2 Fresnel collectors, which provide heat for an ammonia/water ABsorption chiller [24]. The chilled water–glycol is then stored in a 3 m^3 cold storage tank before it is distributed to the wine tanks for cooling purposes. No cost figures have been published for this project.

10.3 The TUI Iberotel Hotel, Turkey

This system provides steam for the laundry and air conditioning for a hotel on the Mediterranean coast in Dalaman, Turkey. It was a collaboration between the TUI Iberotel, DLR and Solitem [25]. It has 40 parabolic trough collectors for process heat and cooling, and has been operational since 2003. The hotel complex consists of 14 villas, each with two floors, with a total of 373 bedrooms.

Figure 10.9 Parabolic trough collectors installed on the ground at the TUI Iberotel in Turkey.

Source: Solitem

Figure 10.10 The steam boiler and double-effect ABsorption chiller at the TUI Iberotel.

Source: Solitem

Table 10.3 shows the system specifications of the solar process heat and cooling. The system layout is shown in Figure 10.11.

Table 10.3 Component specifications for TUI Iberotel solar cooling system

Solar collectors	**Parabolic trough collectors**
Model	Solitem PTC 1800
Total aperture area	360 m^2
Max. working temperature	180 °C (356 °F)
Collector fluid	Pressurised water
Heating capacity	144 kW (DNI 800 W/m^2)
Steam generation	240 kg/h

Chiller	ABsorption
Type	Water/LiBr, double-effect
Model	BROAD
Nominal chilling power output	116 kW_r
Nominal hot water power input	97 kW_{th}
Nominal COP	1.2
Hot water input temperature	180 °C (356 °F)
Chilled water output temperature	6 °C (42.8 °F)
Heat rejection	
Type	Wet cooling tower
Capacity	213 kW_{th}
Heat storage	
Capacity	6 m^3
Fluid	Water
Back-up auxiliary energy source	
Type	LPG fired steam boiler

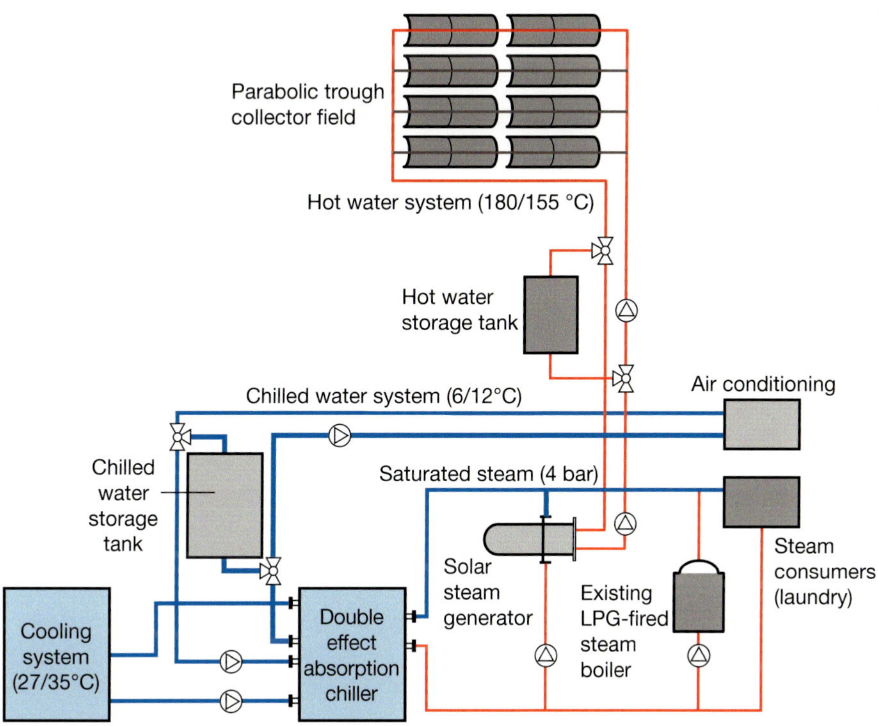

Figure 10.11 Basic system layout of the TUI Iberotel solar cooling system.

Source: Solitem

Parabolic trough collectors with a 360 m^2 collector surface are used to provide pressurised hot water to a solar steam generator, which then provides saturated steam at 4 bar for the laundry and a double-effect water/lithium bromide ABsorption chiller [26]. The power consumption of the solar collector pump is only 0.15 kW$_{el}$. About 6 m^3 of hot water can be stored in a storage tank. The chilled water of the ABsorption chiller is distributed to the air-conditioning system (fan coils).

Figure 10.12 shows the measured performance of the cooling system (heat from fossil and/or solar collectors) for 24 May 2007. The parabolic trough collectors provided about 46 per cent of the total cooling demand on that day. Further performance data is not publicly available yet [26].

Unfortunately, no cost figures are publicly available for this project.

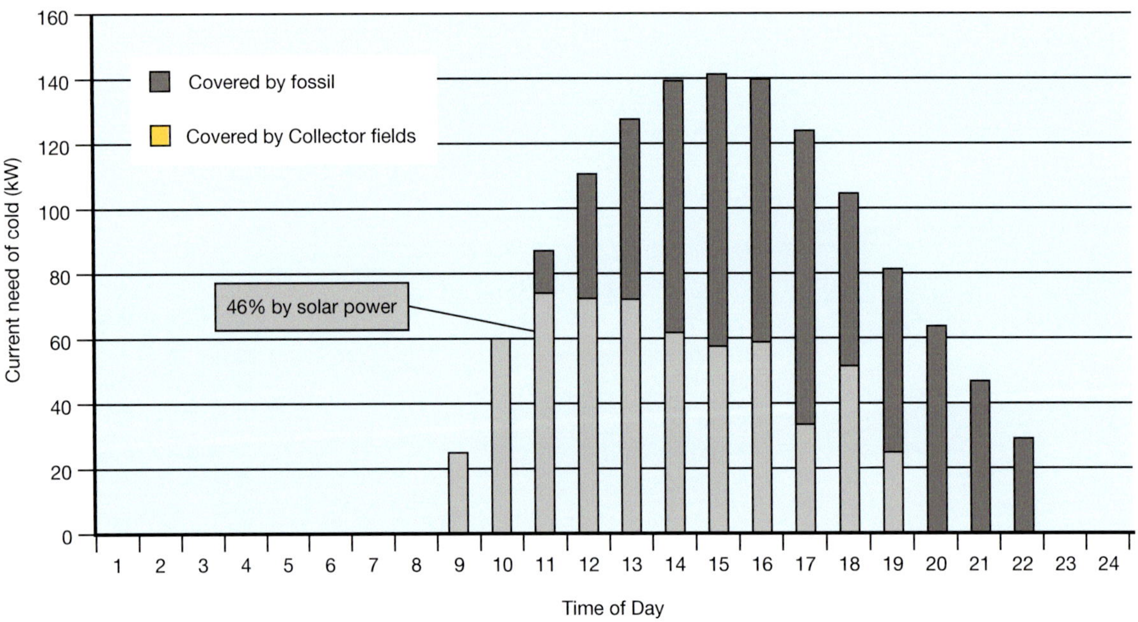

Figure 10.12 Energy composition of chilled water production at the TUI Iberotel.

Source: Solitem

10.4 Ipswich Hospital, Queensland, Australia

Ipswich Hospital, Queensland, Australia has 363 beds and nine floors, and is located in the centre of the city of Brisbane [27]. The project was a collaboration between Queensland Health, Energy Impact, Energy Conservation Solutions and ENERGEX Limited. The system, with 43 parabolic trough collectors, was installed in 2008. In 2013 it was partially removed and no information on the current status was available at the time of printing.

Figure 10.13 Parabolic trough collectors on Ipswich Hospital's multi-storey car park, Queensland, Australia.

Source: Energy Conservation Systems

Figure 10.14 Ipswich Hospital solar collector field.

Source: Energy Conservation Systems

A solar thermal cooling system from Broad was installed at the hospital to support the existing HVAC system. The specifications are given in Table 10.4.

The solar collector field comprises 43 parabolic troughs with 574 m^2 aperture area [27, 28]. It covers 920 m^2 of roof space on the hospital's multi-storey car park. The long parabolic troughs are fixed along the north–south axis, and track the sun along the east–west axis. The solar collector field is controlled by a programmable logic controller (PLC), which monitors radiation levels and moves the troughs to track the sun. The thermal oil in the absorber pipes inside the parabolic troughs is then used to heat water to 175 °C (347 °F). If not required immediately, the water is stored in a 5 m^3 tank. The tank is pressurised with nitrogen and also acts as an expansion device. The storage extends the operational hours of the system into times when there is limited solar radiation, such as at night or when it is cloudy. The hot water is then piped to a double-effect ABsorption chiller (Figure 10.15).

Table 10.4 Component specifications for Ipswich Hospital solar cooling system

Solar collectors	**Parabolic trough collectors**
Model	Broad
Total aperture area	574 m^2
Max. working temperature	175 °C (347 °F)
Collector fluid	Thermal oil
Chiller	**ABsorption**
Type	Water/LiBr, double-effect
Model	BROAD BZH25
Nominal chilling power output	295 kW$_r$
Nominal hot water power input	255 kW$_{th}$
Nominal COP	1.16
Hot water input temperature	175 °C (347 °F)
Chilled water output temperature	6 °C (42.8 °F)
Heat rejection	
Type	Wet cooling tower
Capacity	550 kW$_{th}$
Heat storage	
Capacity	5 m^3
Fluid	Water
Back-up auxiliary energy source	
Type	Gas-fired steam boiler

Figure 10.15 Double-effect ABsorption chiller installed outside of the Ipswich Hospital.

Source: Energy Conservation Systems

Challenges faced during the initial stages of system use included managing:

- high operating pressures with water at high temperatures;
- thermal energy loss experienced in the storage tanks and over the long pipework;
- calibration and control of the solar tracking system.

These challenges have resulted in a longer than expected design and commissioning period. The system has been operating since January 2010. The hospital is fully air-conditioned with a 4.5 MW_r capacity cooling system [27]. The HVAC system accounts for a large amount of the hospital's overall electricity use and peak energy demands. The solar air-conditioning system's collector field has a 255 kW_{th} thermal capacity and can produce around 360 MWh_{th} of thermal energy annually. The system has been designed to provide around 30 per cent of the hospital's peak energy requirements for cooling. The solar cooling system, and other energy-saving initiatives such as a solar hot water system and efficiency upgrades for the HVAC system, avoid around 80 MWh_{el} of electricity use each year. This results in an annual saving of 70 tonnes of CO_2 emissions.

The capital cost of the installed solar cooling system was €692,000 [27]. Annual operation and maintenance cost is €1,030. This covers quarterly inspections and routine cleaning of the solar collectors to maintain efficiency. The investment for the solar collectors were 54.4 per cent (€376,448) of the capital costs, 15.2 per cent (€105,184) for the structural construction including construction drawings and approvals, 3 per cent (€20,760) for solar control box, 1.4 per cent (€9,688) for the hot water storage tank, 15.2 per cent (€105,184) for the double-effect absorption chiller, 5.2 per cent (€35,984) for pipework and 5.8 per cent (€40,136) for pumps, valves and heat exchangers [29]. The structural costs of the solar system were so high because of the difficulties of working on the roof of an operational car park, and because a massive frame capable of withstanding high storm/wind loads had to be installed.

The solar cooling system delivers considerable energy savings during peak demand times as peak solar radiation around midday correlates with peak cooling demands. This also minimises load on the electricity network distribution infrastructure during peak times. The specific costs for the installed solar cooling system are €2,346/kW_r cooling capacity and €1,206/m^2 collector surface.

10.5 Le Bonlait/Best Milk Dairy, Morocco

This system utilises solar process heat from roof-mounted parabolic trough collectors for milk cooling. The system is installed at the Le Bonlait/Bestmilk Dairy in Marrakech. The project was a collaboration between Fraunhofer ISE, POLIMI, Le Bonlait, Abengoa Solar and Tecsol, and has been in operation since 2008 [30].

Table 10.5 shows the specification lists of the system components. The solar collector field comprises 18 parabolic troughs with 70 m^2 aperture area [30]. Thermal oil is used inside the pipes of the parabolic troughs instead of pressurised water. An ammonia/water ABsorption chiller from Robur generates the cold for the milk cooling.

Figure 10.16 Parabolic trough collectors and ammonia/water ABsorption chiller of Dairy Le Bonlait in Marrakech prior to system commissioning.

Source: TECSOL

Table 10.5 Component specifications for Le Bonlait/Bestmilk Dairy solar cooling system

Solar collectors	**Parabolic trough collectors**
Model	Abengoa IST PT-1
Total aperture area	70 m^2
Max. working temperature	180 °C (356 °F)
Collector fluid	Thermo oil
Chiller	**ABsorption**
Type	Ammonia/water, single-effect
Model	Robur ACF 60-00 LB
Nominal chilling power output	12.8 kW$_r$
Nominal hot water power input	23.8 kW$_{th}$
Nominal COP	0.54
Hot water input temperature	180 °C (356 °F)
Chilled water output temperature	–5 °C (23 °F)
Heat rejection	
Type	Directly air-cooled chiller
Capacity	36.6 kW$_{th}$
Cold storage (latent heat)	
Capacity	2 m^3
Fluid	Water–glycol

If cooling is not required immediately, then the cold is stored in a 2 m^3 latent storage tank. The storage tank used here contains plastic balls with a diameter of 99 mm, which are filled with pure water and account for 60 per cent of the storage volume. The rest of the tank volume is filled with brine. If the temperature of the water–glycol mixture in the storage tank drops below 0 °C (32 °F), the water in the balls freezes and thus stores the cold. If the temperature of the brine rises because less cold is produced than needed, the ice melts again and releases the cold. There are no cost figures available for this project.

10.6 Munich Airport canteen, Germany

As part of an innovation project the operator of Munich Airport investigated energy concepts that are sustainable in terms of climate protection and CO_2 savings. One of the resulting pilot plants is the solar cooling system of the airfreight canteen, which has 120 users. The project was a collaboration between Munich Airport and Menerga and has been in operation since February 2009 [31].

Figure 10.17 Looking into the airfreight canteen of Munich Airport makes it clear that the main cooling requirements (cooling loads) are the result of heat entering the canteen via the south-facing windows.

Source: Menerga

Figure 10.18 The flat plate collectors are mounted in front of the airfreight canteen building; the canteen itself is on the top floor.

Source: Menerga

The system is a liquid sorption system combined with 36 flat plate collectors. The air supply is conditioned/dehumidified using lithium chloride solution. The regeneration heat for the salt solution is provided by 75 m^2 flat plate collectors and a 3 m^3 hot water storage tank (Table 10.6). Optimised for summer operation, the solar collectors are aligned mounted at an angle of 30° to the south. For desorption of the lithium chloride solution solar thermal energy can be used in a temperature range of 55–77 °C (131–158 °F). A CHP-driven district heating system is available as a back-up.

Indirect evaporative cooling in the exhaust path cools the air supply in a double cross-flow heat exchanger (Figure 10.19). The nominal air volume flow is 10,000 m^3/h. The proportion of the electrical energy required in the whole system is limited to the power consumption of fans and accessories.

In winter, a heat exchanger with hot water from the solar collectors heats the air supply. For this purpose only a maximum air temperature of 45 °C (113 °F) is needed, which is provided by a powerful heat exchanger in combination with integrated, highly efficient heat recovery from the exhaust air. This allows a low-temperature thermal entry-level use of solar collectors. Excess solar heat is used all year round to pre-heat the hot water for the canteen kitchen. A collector yield of about 400 $kWh_{th}/(m^2\ a)$ is achieved.

Table 10.6 Component specifications for airfreight canteen solar cooling system

Solar collectors	**Flat plate collector**
Model	N.A.
Total aperture area	75 m^2
Max. working temperature	70 °C (158 °F)
Collector fluid	Water–glycol
Chiller	**Liquid sorption**
Type	Lithium chloride
Model	Sorpsolair Type 72
Nominal chilling power output	65 kW_r
Nominal hot water power input	54.2 kW_{th}
Nominal COP	1.2
Hot water input temperature	70 °C (158 °F)
Nominal inlet air temperature	20.3 °C (69 °F)
Heat storage	
Capacity	3 m^3
Fluid	Water
Back-up auxiliary energy source	
Type	District heat from CHP unit

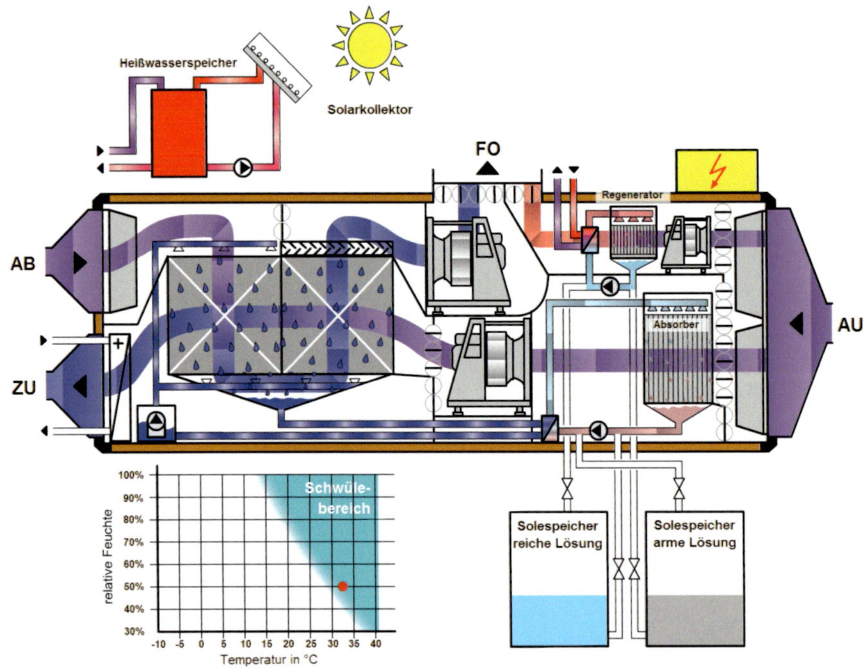

Figure 10.19 The system at Munich Airport is a liquid sorption air-conditioner system with double cross-flow heat exchanger to cool down the air supply via adiabatic cooling of the exhaust air flow.
Heißwasserspeicher: hot water storage
Solarkollektor: solar collector.
Solespeicher: solution storage (brine)
reiche Lösung: diluted brine
arme Lösung: concentrated brine

Source: Menerga

Figure 10.20 Liquid sorption air-conditioner on the roof of the airfreight canteen.

Source: Menerga

The liquid sorption system (Figure 10.19) has an investigated coefficient of performance (COP) of solar input used for the regeneration of the salt solution of about 1.2 (measured average in summer 2009) [31]. Therefore, the costs of evaporative cooling are relatively low. Detailed cost figures of the system are not available.

10.7 Residential building, Wiesloch, Germany

One of the first solar-cooled residential buildings in Europe is located in Wiesloch, near Heidelberg, Germany. The system consists of a small-scale ADsorption chiller combined with 40 in-roof flat plate collectors. The project is the result of collaboration between the building owner family and SolarNext [32, 33], and has been in operational since October 2008. The energy-saving renovation of the building with its 260 m^2 of living space was also carried out. When planning the work, the owner was faced with a tough decision: to buy either a new Porsche or to invest his money in such a unique environment-friendly solar cooling system. He decided for the latter – sorry, Porsche.

Figure 10.21 Roof of the residential building: the in-roof flat plate collectors are the heat source for the small-scale solar cooling system.

Source: Brinkmöller

Figure 10.22 The chillii® Cooling Kit ADsorption chiller in the cellar.

Source: SolarNext

Table 10.7 Component specifications for the Wiesloch residential building solar cooling system

Solar collectors	**Flat plate collector**
Model	BUSO Type ECO-EB
Total aperture area	40 m^2
Max. working temperature	72 °C (162 °F)
Collector fluid	Water–glycol
Chiller	**ADsorption**
Type	Water/silica gel, single-effect
Model	chillii® STC8
Nominal chilling power output	7.5 kW_r
Nominal hot water power input	13.4 kW_{th}
Nominal COP	0.56
Hot water input temperature	72 °C (162 °F)
Chilled water output temperature	15 °C (59 °F)
Heat rejection	
Type	Dry cooler with adiabatic option
Capacity	21 kW_{th}
Heat storage	
Capacity	2 m^3
Fluid	Water
Back-up auxiliary energy source	
Type	Oil burner

The system is one of the first standardised solar cooling kits, a chilli® Cooling Kit STC8 from SolarNext AG. It consists of a 7.5 kW_r ADsorption chiller (see Figure 10.23) with an effective waste heat rejection of 21 kW_{th}. This is done using an electrically high-efficient dry cooler with optional adiabatic water spraying and a system controller. The driving heat for the chiller is provided by 40 m^2 of in-roof flat plate collectors, which are installed with a 25° inclination to south-southeast and an oil burner back-up. A 2,000 litre hot water storage tank is used as a buffer. An air-handling unit is used for the cold distribution. The system layout is shown in Figure 10.24.

The capital cost of the installed system, which provides domestic hot water, heating and cooling, was €60,400 [32]. A subsidy of €8,400 for the solar collectors was obtained from the German government office BAFA through its

Figure 10.23 Dry cooler installed in the garden besides the building. The adiabatic water spray pipe can be seen underneath the cooler. It is only operated during periods of high ambient temperature.

Source: SolarNext

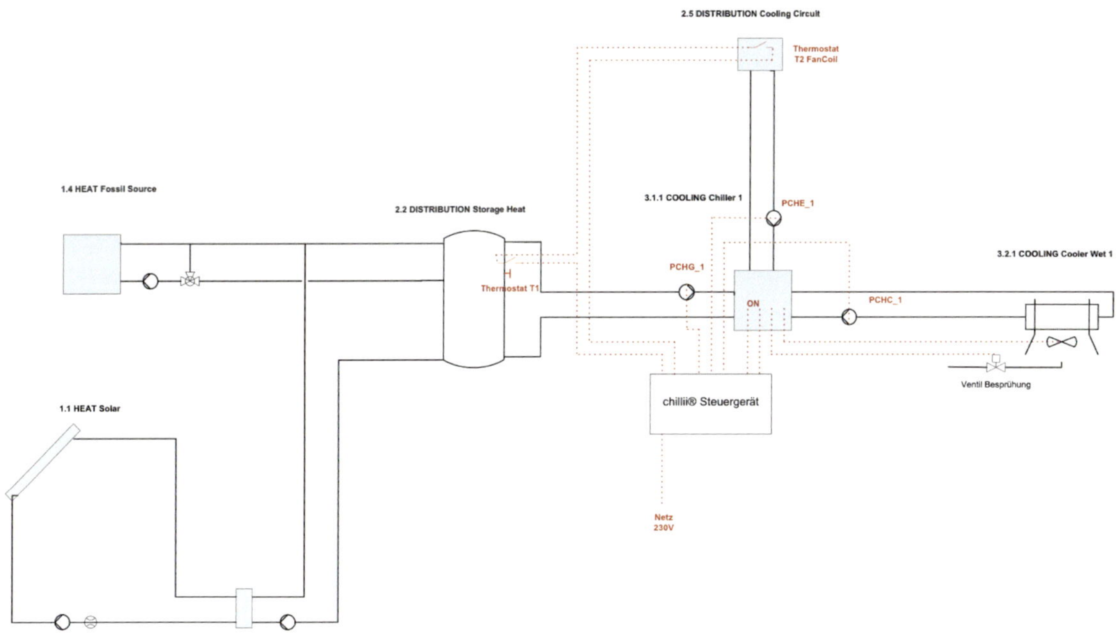

Figure 10.24 System layout of the solar cooling kit installed at the residential building.

Source: SolarNext

market incentive programme (MAP), making the net total cost €52,000. The solar collectors and heat storage tank cost €13,500 (22 per cent of total costs); the oil burner cost €6,700 (11 per cent); the chilli® Cooling Kit cost €17,000 (28 per cent); the cold distribution cost €13,200 (22 per cent) and the actual installation cost €10,000 (17 per cent). The specific costs for the installed solar cooling kit, including subsidies, were €6,933/kW_r for the cooling capacity and €1,300/m^2 for the collectors.

10.8 Charlestown Square Shopping Centre, Newcastle, Australia

The solar cooling system that has been installed on the Charlestown Square Shopping Centre in Charlestown, NSW, Australia consists of 12 NEP solar PolyTrough 1200 collectors, powering a double-effect 233 kW_r ABsorption chiller providing chilled water to a central chilled water network. The system was installed in 2010 and was funded by the building owner, GPT, and a NSW government grant. The shopping centre has 88,000 m^2 floor space and over 270 stores.

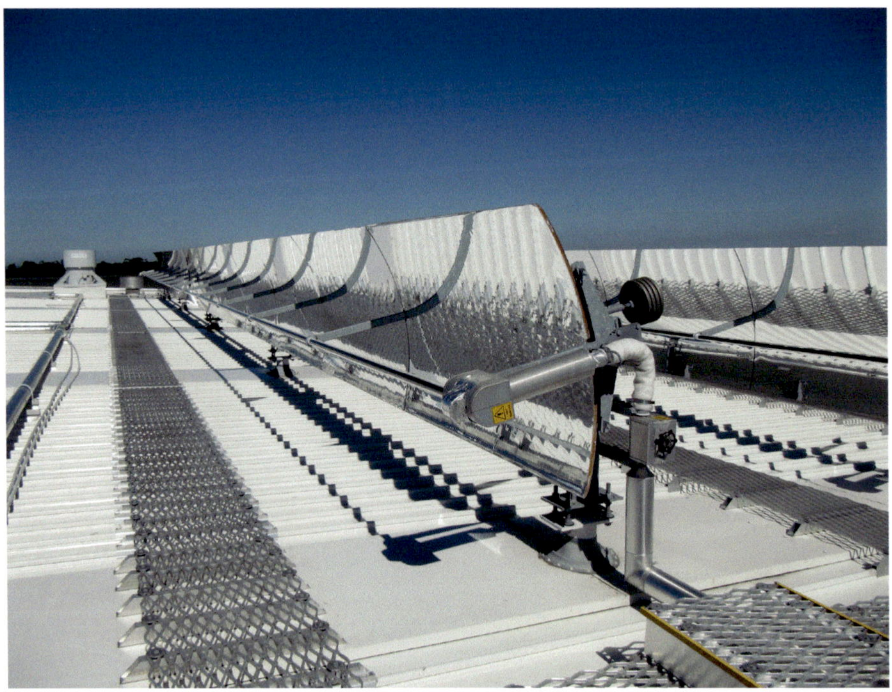

Figure 10.25 The Charlestown Square Shopping Centre NEP collectors in stow position. An advantage of concentrated collectors is that if the system is not in operation the collector surface can be turned away from the sun to prevent stagnation.

Source: Solem Consulting

Figure 10.26 Pressurised water-driven double-effect ABsorption chiller installed in the technical room at the shopping centre.

Source: Solem Consulting

A double-effect ABsorption chiller combined with 350 m^2 of parabolic trough collectors (Table 10.8) supplies chilled water for a much larger conventional conditioning plant consisting of several conventional compressor chillers and a CHP waste-heat-driven, single-effect ABsorption chiller. The solar cooling system is expected to produce an estimated 230 kW$_r$ of cooling at peak output and will reduce conventional cooling electricity use by 77,000 kWh$_{el}$/a. The system is designed to cut down the electricity peak loads of the conventional conditioning plant during peak daytime. The system layout is shown in Figure 10.27.

The project was funded by the building owner and by a government grant from the Renewable Energy Development Program. The total cost of the project was AU$1,061,000 (about €727,680). Relative costs of the system's components can be seen in Figure 10.28.

One of the lessons learned during this project is that it should be clear who is responsible for overall system design, not just the design, selection and sizing of the individual system components. This may be achieved via a turn-key provider or via an engineering consulting firm that fully understands the system as a whole and can correctly design, size and select the system components so that they function well together.

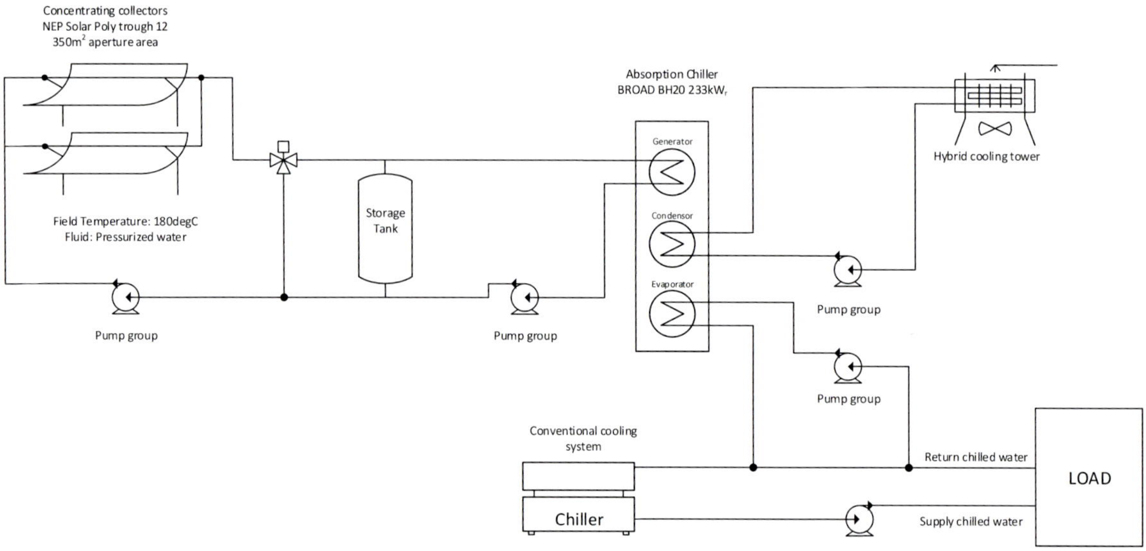

Figure 10.27 System layout of the solar cooling system installed at the shopping centre.

Source: Solem Consulting

Table 10.8 Component specifications for the shopping mall solar cooling system

Solar collectors	**Parabolic trough collectors**
Model	NEP PolyTrough 1200
Total aperture area	350 m^2
Max. working temperature	180 °C (356 °F)
Collector fluid	Pressurised water
Chiller	**ABsorption**
Type	Water/lithium bromide, double-effect
Model	Broad BH20
Nominal chilling power output	233 kW$_r$
Nominal hot water power input	160 kW$_{th}$
Nominal COP	1.4
Hot water input temperature	180 °C (356 °F)
Chilled water output temperature	7 °C (44.6 °F)
Heat rejection	
Type	Hybrid cooling tower
Capacity	393 kW$_{th}$

Heat storage	
Capacity	2 m^3
Fluid	Water
Back-up auxiliary energy source	
Type	None – conventional plant for building can handle cooling demand when solar not available

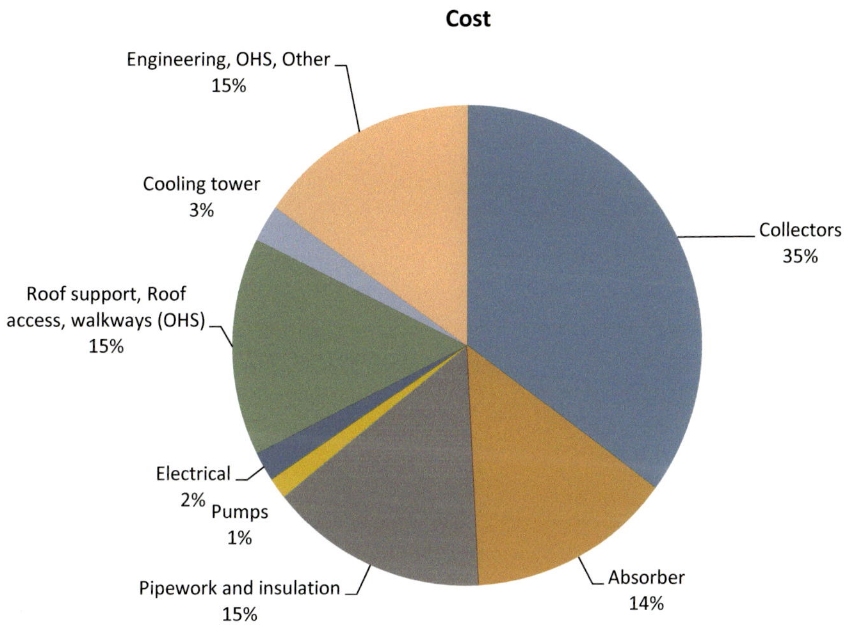

Figure 10.28 Cost distribution of shopping centre solar cooling system.

Source: Solem Consulting

11 Glossary

Absorber area – the area of a solar thermal collector where radiation energy is converted into thermal energy

ABsorption – the incorporation of a substance in one phase state into another substance in a different phase state (e.g. gases being absorbed by a liquid). This involves the whole volume of the liquid

Absorptivity – the fraction of radiation absorbed at a specific wavelength

ADsorption – the physical bonding of a gaseous or liquid substance onto the surface of another substance of a different phase (e.g. vapour adsorbed to a solid surface). This process involves the surface only

Air-based systems – open cold distribution systems using air as the heat transfer medium in air ducts to supply conditioned air to the rooms and to collect the exhaust air from the building

Annuity method – method for investment calculations using the annuity factor. It refers to a time series of constant cash flows over multiple years. Annuity is calculated based on the first year, assuming constant cash flows for all following years

Aperture area – the area of a solar thermal collector that collects the sun's radiation energy (typically larger than the absorber area)

Aquifer – a layer of permeable rock, gravel, sand or silt that contains groundwater. Extraction of the water is usually via boreholes

Chilled water – water used for process cooling or air-conditioning purposes; here, the water that flows in the circuit of an AB-/ADsorption chiller with the lowest temperature

CHP – combined heat and power generation, also known as cogeneration. Refers to the process of simultaneous generation and use of heat and power, typically from a cogeneration unit. Heat is released from all thermal power generation processes (e.g. coal-fired power plants); however, it is usually ejected into the environment (as steam) and therefore lost. CHP units capture most of this heat, which increases the fuel utilisation efficiency

Cogeneration – see CHP

Concentration ratio – the ratio of aperture to absorber area of a solar collector

Cooling capacity – the cooling power of a thermal chiller, typically referring to the energy removed in the chilled water circuit

Cooling degree days – Cooling degree days (CDDs) are determined by the difference between the average daily temperature and the comfort-level temperature. If cooling is being considered to a temperature comfort level of 24 °C, and if the average temperature for a day was 27 °C, then cooling equivalent to 3 °C (3 CDDs) would be required to maintain a temperature of 24 °C for that day. However, if the average temperature was 21 °C, then no cooling would be required, so the number of CDDs for that day would be zero. Note that degrees refers to the local temperature system (e.g. Celsius or Fahrenheit)

Cooling water – water used to reject the heat from a building, a thermal cooling process; here, the water that flows in the circuit of an AB-/ADsorption chiller with the medium temperature

Dehumidification – dehumidification of air through the dew point

Density – the mass unit per volume of a substance

Dew point temperature – the temperature below which the water vapour in air at constant air pressure is condensed into water in the same amount with which it was evaporated

Diffuse radiation – solar radiation that interacts with the atmosphere before reaching the Earth's surface. Diffuse radiation is non-directional. It is caused by scattering of light at molecules in the atmosphere, reflection and absorption/re-emission

Direct radiation – solar radiation that does not interact with the atmosphere before reaching the Earth's surface. Direct radiation is directional and comes straight from the sun

Discount rate – the rate of interest used in NPV calculations. It reflects the rate of return which could be earned in the global financial markets for an investment with similar risk

District heating/cooling – a process whereby heat or cold is distributed from a central plant via insulated pipes to residential and/or commercial buildings to provide them with heating or cooling

DN pipe size – diameter nominal; a measure for pipe diameter according to EN ISO 6708

DNI – direct normal irradiation; the direct radiation falling orthogonally onto a collector surface

Drain-back system – a solar thermal system with a separate heat carrier fluid catchment tank. The heat carrier fluid contained in the collectors and piping above the catchment tank drains into this tank if the solar circuit pump is stopped. This protects the fluid from damage due to frost and stagnation overheating

Dry cooler – a dry cooler is a heat exchanger with a fan on the top to dissipate heat from the chiller to the ambient

Emissivity – the relative ability of a material surface to emit energy by radiation. It is expressed as the ratio of energy radiated by a material at a given temperature to energy radiated by a black body at the same temperature

Evaporative cooling – the process of cooling a water body or flow via the partial evaporation of some of the water. Heat required for the evaporation of water is removed from the body of water itself, hence cooling it

Feed-in tariff (FiT) – a financial mechanism where renewable energy electricity generators are paid a fixed price for the electricity exported to the grid. Typically contracts are long term, ensuring manageable risks for investors in renewable energy technologies

Free cooling – use of an installed heat rejection system (see wet cooling tower or dry cooler) simply using the temperature difference to low ambient air temperatures conditions

Gross collector area – the total area of a solar thermal collector required by the complete solar installation on a roof or the ground. Gross collector area includes space required for access and peripherals

Heat convection – a heat transfer process based on the movement of liquids or gases

Heating degree days – heating degree days (HDDs) are determined by the difference between the average daily temperature and the comfort-level temperature. For example, if heating is being considered to a temperature comfort level of 18 °C, and the average daily temperature for a particular location was 14 °C, then heating equivalent to 4 °C (4 HDDs) would be required to maintain a temperature of 18 °C for that day. However, if the average daily temperature was 20 °C then no heating would be required, so the number of HDDs for that day would be zero. Note that degrees refers to the local temperature system (e.g. Celsius or Fahrenheit)

Hot water – water used to drive a thermal cooling or air-conditioning process; here, the water that flows in the circuit of an AB-/ADsorption chiller with the highest temperature

HVAC – heating, ventilation and air-conditioning

IAM – incident angle modifier, which accounts for the influence of non-perpendicular incident radiation at incidence angle Θ in relation to normal incidence radiation ($\Theta = 0$)

Incidence angle – the angle between a light ray and the collector glass cover

Inverter – a device that converts direct current (DC) output from a photovoltaic (PV) solar panel into alternating current (AC)

Latent heat – the energy change of a substance when changing phases at constant temperature (latent heat of fusion for melting and freezing; latent heat of vaporisation for boiling and condensing)

Latent heat storage – storage tanks using a phase change for storing energy (see Latent heat)

Load profile – the distribution of heating and cooling requirements of a building over time. The load profile can be expressed for daily, weekly, monthly and annual time periods

NPS pipe size – nominal pipe size; a measure for pipe diameter according to the American National Standards Institute (ANSI)

NPV method – method for investment calculations using the net present value (NPV). It refers to a time series of variable cash flows, i.e. changing cash inflows and outflows over multiple years. NPV is the present value of the difference between cash inflows and cash outflows, taking inflation and returns into account

Optical collector efficiency – the maximum efficiency of a solar collector. The optical efficiency denotes the optical losses of a collector, consisting of reflection and absorption in the glass cover as well as reflection from the absorber sheet. It is the point of the efficiency curve where the mean collector temperature equals the environment temperature

Panel – a PV panel (module) is an assembly of photovoltaic cells. It provides direct current (DC) when exposed to sunlight

PCM – phase change material; a material that is used in latent heat storage applications. PCMs are chosen according to the temperature level of the storage application

PLC – programmable logic controller; a hardware component of a solar cooling system used for system monitoring and control

Reflectivity – the fraction of radiation that is reflected at a surface

RT – refrigeration ton, unit for the cooling capacity of chillers. 1 RT = 3,517 W_r

Sensible heat – sensible heat refers to the energy change of a substance that has a change of temperature as its only cause

Solar fraction – the fraction of the total annual hot water, heating or cooling requirements of a building supplied by a solar system

Solar yield – the useful thermal energy from solar thermal collectors; usually calculated per year

Sorption – collective term for AB- and ADsorption processes; see AB-/ADsorption for details

Specific heat capacity – the quantity of heat required to raise the temperature of a given mass of substance by a given amount

Storage density – a measure of the ability of a substance to store energy per volume or mass unit

Thermal conductivity – the ability of a substance to conduct heat

Thermal (ground) probe – a U-shaped pipe that is inserted into sealed vertical or inclined borings. Water is pumped through the pipe which then acts as a heat exchanger and either absorbs the heat from the earth or releases heat into the earth

VSD – variable speed drive; a device that allows speed control of electric motors, e.g. for pumps

Water-based systems – closed cold distribution systems using chilled water as a heat transfer fluid in pipes. Heat is taken up from the space to be cooled by means of different water-to-air heat exchangers

Wet bulb temperature – for given ambient conditions, the wet bulb temperature is the lowest temperature that can be reached by the evaporation of water only

Wet cooling tower – heat exchangers that are used to dissipate heat loads from the chiller or the building (see Free cooling) by part evaporation of the cooling water to the ambient

Additional terms can be found in the following two internet glossaries:

US Department of Energy glossary of energy-related terms: http://energy.gov/energybasics/articles/glossary-energy-related-terms

Mother Earth News renewable energy glossary: www.motherearthnews.com/renewable-energy/energy-glossary-zl0z1001zken.aspx#axzz2KQyAZzjG

12

Sources of further information

12.1 Units

Table 12.1 contains metric to imperial conversions for selected parameters.

Table 12.1 Conversion table of metric to imperial units

Name	Metric unit	Imperial unit	Name of imperial unit	Conversion factor
Temperature	°C	°F	Degree Fahrenheit	[°F] = [°C] · 1.8 + 32
Power	kW	Btu/hr	British thermal units per hour	1 kilowatt = 3 412.14 Btu/hour [I.T.]
Energy	kJ	Btu	British thermal units	1 kilojoule = 0.948 Btu
Energy	kWh	Btu	British thermal units	1 kilowatt hour = 3 412.14 Btu
Mass	kg	lb	Pound	1 kilogram = 2.204 lb
Volume	Litre	gal	Gallon (liquid US)	1 litre = 0.264 gallon [US, liquid]
Volume	Litre	cu.in.	Cubic inch	1 litre = 61.024 cu.in.
Area	m^2	sq.ft	Square foot	1 square metre = 10.764 square foot
Density	kg/l	lb/cu. ft	Pounds per cubic foot	1 kilogram/litre = 62.428 lb/cu.ft
Thermal conductivity	W/(m K)	Btu/(hr ft °F)	British thermal units per (hour foot Deg F)	1 W/(m K) = 0.57778 Btu/(hr ft F)
Spec. heat capacity	kJ/(kg K)	Btu/(°F lb)	British thermal units per (pound Deg F)	1 kilojoule/(kilogram K) = 0.2388 BTU/(lb °F)

12.2 Country codes, policy, regulations and incentives

Table 12.2 gives an overview of existing country codes, policy, regulations and incentives (status 05/2013), which are directly linked to solar cooling.

Table 12.2 Overview on country codes, policy, regulations and incentives for solar cooling

Country	Title	Contact organisation	General information
Austria	Solarthermie – solare Großanlagen	Klima- und Energiefonds of Austrian federal government, www.kimafonds.gv.at	Grants for solar thermal installations >100 m^2 including solar cooling, budget limited to €750,000. Systems should also prepare DHW and heating of the buildings too. Monitoring is obligatory.
Austria	FI – Umweltförderung Inland	Kommunal Kredit Public Consulting, www.kpc.at	Grants for solar thermal installation, depending on achievable CO_2 savings and grant for sorption chiller and further components depending on installed kW_r chiller capacity. There is a minimum investment for the solar thermal installation of €10,000 and 4.5 kW_r cooling capacity.
France	Emergence program	ADEME, www2.ademe.fr ENERPLAN, www.enerplan.asso.fr	Grants (pre-feasibility phase 50 per cent funding, project financing phase up to 80 per cent funding) will be insured by an overall thermal useful energy per collector area unit of 350 kWh/m^2 a, minimum annual electrical COP_{el} of 5 and the installation has to be monitored for at least the first year.
Germany	Erneuerbare-Energien-Wärmegesetz EEWärmeG	Bundesministerium für Umwelt, Naturschutz und Reaktorsicherheit BMU, www.bmu.de	EEWärmeG defines the use of renewable energies for new residential, non-residential buildings and existing public buildings (based on EU directive 2009/28/EC). When using solar thermal collectors a share of 15 per cent renewable energies are sufficient to fulfil the duty.
Germany	Marktanreizprogramm MAP	Bundesamt für Wirtschaft und Ausfuhrkontrolle BAFA, www.bafa.de	BAFA offers €180/m^2 for collector areas between 20 m^2 and 100 m^2 for solar thermal cooling, and also for new buildings (MFH, commercial).
Germany	Marktanreizprogramm MAP	Kreditanstalt für Wiederaufbau KfW, www.kfw.de	KfW offers a subsidy equal to up to 50 per cent of investment for solar cooling with collector areas between 40 m^2 and 100 m^2.
Germany	Gewerbliche Klima- und Kälteanlagen	Bundesamt für Wirtschaft und Ausfuhrkontrolle BAFA, www.bafa.de	The BAFA programme for promotion of efficient cooling systems in industry offers a subsidy of 25 per cent of net investment on sorption technology >50 kW_r, or 35 per cent if ammonia is used as the refrigerant.
Italy	Bando solare termico	Regione Lombardia, www.regione.lombardia.it	Subsidies for solar thermal systems >50 m^2 for public buildings. An overall thermal useful energy per collector area m^2 gives access to a certain subsidy per m^2. An extra subsidy is given on the basis of kW_r chiller capacity. System performance is measured and the subsidy depends on reaching the declared value.
USA	Energy Smart New Construction Program	New York State Energy Research and Development Authority, www.nyserda.ny.gov	NYSERDA encourages the use of renewable energies for agricultural, commercial, industrial/manufacturing, institutional and multi-family buildings. Incentives of 50–75 per cent of incremental costs depending on type of project, maximum incentive up to US$825,000 for upstate residents and US$1.575 million for Con Edison customers (www.coned.com).

12.3 Books

A list of recommended literature for further reading can be found as follows:

- Eicker, U. (2003) *Solar Technologies for Buildings.* John Wiley & Sons, ISBN 0-471-48637-X
- Eicker, U. (2009) *Low Energy Cooling for Sustainable Buildings*, John Wiley & Sons, ISBN 978-0-470-69744-3
- Final reports of Task 25, *Solar Assisted Air-Conditioning of Buildings, Solar Heating and Cooling Programme, International Energy Agency (IEA).* Available for download from: http://iea-shc-task38.org/task-25. Last accessed 27 February 2013
- Final reports of Task 38, *Solar Air-Conditioning and Refrigeration, Solar Heating and Cooling Programme, International Energy Agency (IEA).* Available for download from: http://iea-shc-task38.org. Last accessed 27 February 2013
- Henning, H.M. (2004) *Solar-Assisted Air-Conditioning in Buildings: A Handbook for Planners*, Springer-Verlag Wien, ISBN 3-111-00647-8
- Henning, H.M., Motta, M. and Mugnier, D. (2013) *Solar Cooling Handbook: A Guide to Solar Assisted Cooling and Dehumidification Processes*, Springer-Verlag Wien, ISBN 978-3-709-10841-3
- Henning, H.M., *et al.* (2009) *Kühlen und Klimatisieren mit Wärme* (in German), BINE Informationsdienst, Fraunhofer IRB Verlag, ISBN 978-3-8167-8324-4
- *OTTI Proceedings of the International Conference Solar Air-Conditioning*, available from: www.renewable-energy-books.com. Last accessed 8 May 2013
- Saha, B.B., Chakraborty, A. and Choon Ng, K. (2011) *Innovative Materials for Processes in Energy Systems for Fuel Cells, Heat Pumps and Sorption Systems*, Research Publishing Singapore, ISBN: 978-981-08-7614-2
- Weiss, W. (2003). *Solar Heating Systems for Houses: A Design Handbook for Solar Combisystems*, Earthscan UK, ISBN: 978-1-902-91646-0

12.4 Courses

Design courses for solar cooling systems are being offered through various organisations worldwide.

Table 12.3 Global overview on design courses for solar cooling systems

Name of supplier	Country	Course name	Course duration	Content	Website
Solem Consulting	Germany/ Australia	Solar cooling seminar	1 day to 2 weeks, depending on content	System design, control, economic calculations, modelling and simulation. Courses are tailored to the client's requirements. Please contact the course supplier for an individual offer.	www.solem-consulting.com
AEE INTEC	Austria	SOLAIR	Please contact the national course supplier for an individual offer	Introduction, basics, pre-design, design, economics and environment, installed systems and success stories	www.solair-project.eu
Ambiente Italia	Italy				
Provincia di Lecce	Italy				
Politecnico di Milano	Italy				
CRES	Greece				
Fraunhofer Institut	Germany				
INETI	Portugal				
EVE	Spain				
AIGUASOL	Spain				
TECSOL	France				
REHVA	The Netherlands				
University of Ljubljana	Slovenia				
CED Engineering.com	USA	An introduction to solar cooling systems (R02-002)	Online course	Absorption cooling, Rankine cycle, desiccant cooling, other cooling methods, estimating system size, system controls, piping, collectors, other considerations	www.cedengineering.com
CSIRO	Australia	Solar cooling crash course	Please contact the course supplier for an individual offer	Please contact the course supplier for an individual offer	www.csiro.au

Green Engineers	Germany	Solar cooling workshop	Please contact the course supplier for an individual offer	Please contact the course supplier for an individual offer	green-engineers.de
IZT	Germany	Powerado-Plus, Modul Solare Kühlung (13)	2 days	Basics, technology of thermally driven chillers, system integration, overview of small-scale suppliers, design and concept, economics, experiences with pilot plants, examples	http://projekte.izt.de
Thermosol Consulting	Canada	Solar cooling workshop	Please contact the course supplier for an individual offer	Please contact the course supplier for an individual offer	www.thermosolconsulting.com

12.5 Organisations

A list of organisations related to solar cooling can be found as follows:

- ASHRAE, American Society of Heating, Refrigerating and Air-Conditioning Engineers, www.ashrae.org
- AusSCIG, Australian Solar Cooling Interest Group, www.ausscig.org
- Green Chiller Association for Sorption Cooling e.V., www.greenchiller.eu
- IEA, International Energy Agency, www.iea.org
- IEA-SHC, Solar Heating and Cooling Program, www.iea-shc.org
- IIF/IIR, International Institute of Refrigeration, www.iifiir.org
- US Solar Heating and Cooling (SHC) Alliance, www.seia.org/

12.6 Solar radiation databases

Solar radiation databases for the design of solar cooling systems are being offered through various companies and organisations worldwide. The following list does not claim to be exhaustive

- AccuRate Software, Australian weather data, www.csiro.au/en/Organisation-Structure/Flagships/Energy-Transformed-Flagship/AccuRate.aspx#a4
- EnergyPlus Energy Simulation Software 8.0.0, Add-On Weather Data, http://apps1.eere.energy.gov/buildings/energyplus/weatherdata_about.cfm
- Meteonorm Software 7.0, http://meteonorm.com
- NASA Surface Meteorology and Solar Energy, https://eosweb.larc.nasa.gov
- Satel-Light, The European Database of Daylight and Solar Radiation, www.satel-light.com
- Solar Pathfinder, Solar Radiation Data for USA, www.solarpathfinder.com/solar_radiation

12.7 Suppliers

Table 12.4 offers an overview on suppliers of thermal sorption chillers, concentrating collectors and desiccant-evaporative systems (DECs). Suppliers for other collector types and components are too numerous to display here. The following does not claim to be exhaustive.

Table 12.4 Overview on component suppliers for solar cooling systems

	Name of supplier	Country	Technology	Website
Thermal sorption chiller	AGO	Germany	ABsorption chillers	www.ago.ag
	Broad	China	ABsorption chillers	www.broad.com
	Carrier	USA	ABsorption chillers	www.commercial.carrier.com
	Daikin McQuay	USA	ABsorption chillers	www.daikinmcquay.com
	EAW	Germany	ABsorption chillers	www.eaw-energieanlagenbau.de
	Energy Concepts LLC	USA	ABsorption chillers	www.energy-concepts.com
	Fischer Eco Solutions	Germany	ABsorption chillers	www.fischer-group.com
	Hitachi	Japan	ABsorption chillers	www.hitachi-pt.com
	InvenSor	Germany	ADsorption chillers	www.invensor.com
	Jiangsu Huineng	China	ABsorption chillers	www.tynkt.com
	Kawasaki	Japan	ABsorption chillers	www.khi.co.jp
	Köhler Industries	Germany	ABsorption chillers	www.koehler-industries.de
	Mayekawa	Japan	ADsorption chillers	www.mayekawa.com
	Mitsubishi Plastics	Japan	ADsorption chillers	www.mpi.co.jp
	HIJC	USA	ADsorption chillers	www.greenchiller.biz
	Pink	Austria	ABsorption chillers	www.pink.co.at
	Robur	Italy	ABsorption chillers	www.robur.it
	Sakura	Japan	ABsorption chillers	www.sakura-aircon.com
	Shuangliang	China	ABsorption chillers	www.shuangliang.com
	Tranter SolarIce	Germany	ABsorption chillers	www.solarice.de
	SolabCool	The Netherlands	ADsorption chillers	http://solabcool.com
	SorTech	Germany	ADsorption chillers	www.sortech.de
	Thermax	India	ABsorption chillers	www.thermaxindia.com
	Yazaki	Japan	ABsorption chillers	www.yazaki-airconditioning.com
	York	USA	ABsorption chillers	www.johnsoncontrols.com

	Name of supplier	Country	Technology	Website
Concentrating collectors	Areva	France	Compact linear Fresnel	www.areva.com
	Chromasun	USA	Micro linear Fresnel	http://chromasun.com
	Novatec Solar	Germany	Compact linear Fresnel	http://novatecsolar.com
	Industrial Solar	Germany	Compact linear Fresnel	www.industrial-solar.de
	NEP Solar	Switzerland/ Australia	Parabolic trough	http://nep-solar.com
	Solarlite	Germany	Parabolic trough	www.solarlite.de
	Smirro	Germany	Parabolic trough	www.smirro.de
	Soltigua	Italy	Parabolic trough/ linear Fresnel	www.soltigua.com
	Shap	Italy	Parabolic trough	www.shap.it
	Sopogy	USA	Parabolic trough	http://sopogy.com
	Absolicon	Sweden	Parabolic trough	www.absolicon.com
	Heliodynamics	Canada	Parabolic trough	www.heliodynamics.com
	IT Collect	Germany	Parabolic trough	www.itcollect.de
	Pro Target	Germany	Parabolic trough	www.protarget-ag.de
	SRB Energy	Switzerland	CPC	www.srbenergy.com
	Thermax	India	Parabolic trough	www.thermaxindia.com
	SLT Energy	India	Parabolic trough	www.sltenergy.com
Desiccant-evaporative systems (DECs)	Menerga	Germany	Liquid sorption	www.menerga.com
	Munters	Sweden	Solid sorption	www.munters.com
	Sigle und Epple	Germany	Solid sorption	www.siegleundepple.de
	Wolf Geisenfeld	Germany	Solid sorption	www.wolf-geisenfeld.de
	L-DCS	Germany	Liquid sorption	www.l-dcs.com
	AIL Research	USA	Liquid sorption	www.ailr.com

References

[1] Stahl, M., 2012 war kein gutes Jahr für Raumklimageräte, www.cci-dialog.de, 2013.

[2] Mouchot, A., *Die Sonnenwärme und ihre industriellen Anwendungen*, Olynthus-Verlag, 1987.

[3] Grossman, G., Solar-powered systems for cooling, dehumidification and air-conditioning, *Solar Energy* Vol. 72, No. 1, pp. 53–62, 2002.

[4] http://wardsauto.com/sales-amp-marketing/world-vehicle-sales-surpass-80-million-2012. Last accessed 5 July 2013.

[5] Stapleton, G. and Neill, S. (2011). *Grid-Connected Solar Electric Systems: The Earthscan Expert Handbook for Planning, Design and Installation*, Earthscan, 2011.

[6] Eicker, U., *Low Energy Cooling for Sustainable Buildings*. John Wiley & Sons, 2009.

[7] www.solarbuzz.com/facts-and-figures/retail-price-environment/module-prices. Last accessed 19 September 2013.

[8] Quaschning, V., *Erneuerbare Energien und Klimaschutz: Hintergründe – Techniken und Planung – Ökonomie und Ökologie – Energiewende*, Carl Hanser Verlag, 2013.

[9] ITW Stuttgart, Germany. Personal communication, 2013.

[10] Laughton, C., *Solar Domestic Water Heating: The Earthscan Expert Handbook for Planning, Design and Installation*, Earthscan, 2010.

[11] Posladek, G., An investigation into using free cooling and community heating to reduce data centre energy consumption. MSc Thesis, University of Strathclyde, Glasgow, UK, 2008.

[12] TECSOL, Check-List Method for the Selection and the Success in the Integration of a Solar Cooling System in Buildings, International Energy Agency (IEA), 2011. Available for download from: www.tecsol.fr/checklist. Last accessed 12 May 2013.

[13] Kuehn, A. and Ziegler, F., Operational results of a 10 kW absorption chiller and adaptation of the characteristic equation. *Proceedings of the 1st International Conference Solar Air Conditioning*, OTTI, Germany, 2005

[14] Kuehn, A., Meyer, T. and Ziegler, F., Operational results of a 10 kW absorption chiller in heat pump mode. *Proceedings of the 9th International Energy Agency Heat Pump Conference*, Zürich, Switzerland, 2008.

[15] Franzke, U., Design and control of SDEC systems. Contribution to IEA-SHC Task 38 report, *Solar Cooling System Design and Control*, ILK Dresden, Germany, 2009.

[16] http://archive.iea-shc.org/publications/downloads/IEA-Task38-Report_A5_final.pdf. Last accessed 5 July 2013.

[17] Weiss, W., *Solar Heating Systems for Houses: A Design Handbook for Solar Combisystems*, Earthscan UK, 2003.

[18] Hausner, R. and Fink, C., Stagnation behaviour of solar thermal systems. *Proceedings of EUROSUN Conference*, 19–22 June 2000, Copenhagen, Denmark.

[19] www.solair-project.eu/123.0.html. Last accessed 5 July 2013.

[20] Greer, D., Songs in the Key of Green, New York Construction, http://newyork.construction.com/features/archive/2009/05_F2a_SolarSonata.asp. Last accessed 11 January 2011.

[21] Epp, B., *Steam and Cooling: No Problem for a Solar Thermal System*, www.solarthermalworls.org/node/432. Last accessed 11 January 2011.

[22] Henkel, T., Solar ABsorption Cooling and Heating Systems. *Proceedings of the Solar San Antonio Solar Cooling Workshop*, 23 February 2010, San Antonio, USA.

[23] Ounalli, A. and Nefzi, K., Project MEDISCO. *Proceedings of the International Conference Berbi 2008: Solar Cooling Session*, 5 May 2008, Perpignan, France.

[24] Motta, M., Solar Cooling II, SOLNET course, Uni Kassel, Germany, March 2008.

[25] Lokurlu, A., Richarts, F. and Krüger, D., High efficient utilisation of solar energy with newly developed parabolic trough collectors (SOLITEM PTC) for chilling and steam production in a hotel at the Mediterranean coast of Turkey. *International Journal of Energy Technology and Policy*, Vol. 3, No. 1/2, 2005, pp. 137–146.

[26] Lokurlu, A., Advanced Solar Cooling System. *Proceedings of the 7th Deutsch-Portugiesische Symposium Erneuerbaren Energien*, 9 November 2010, Lisbon, Portugal.

[27] DEEDI, Ipswich Hospital, Queensland: Solar thermal air-conditioning, Cleanenergy, June 2010.

[28] Kohlenbach, P. and Dennis, M., Solar Cooling in Australia: The Future of Air-Conditioning? *Proceedings of the 9th IIR Gustav Lorentzen Conference*, 12–14 April 2010, Sydney, Australia.

[29] Braunsberger, A., Ipswich Hospital: Energy Efficiency Project Solar Cooling System. *Proceedings of the ausSCIG Solar Cooling 2011 Conference*, 16 March 2011 Canberra, Australia.

[30] Klempert, O., Process cooling for Africa's food industry. *Sun & Wind Energy*, Vol. 9, No. 2, 2011, pp. 74–76.

[31] Menerga, Kühlen mit offenen Sorptionssystemen, cci Technik hoch zwei, No. 4, 2010, pp. 22–25.

[32] Brinkmöller, B., Solare Kühlung im Zweifamilienhaus. *Proceedings of the Seminar Solare Klimatisierung, Energie der kurzen Wege*, 16 October 2008, TP Conference Center Heidelberg, Germany.

[33] Jakob, U., Performance and costs of low capacity solar cooling systems. *Proceedings of the CEP09 International Symposium on Solar and Renewable Cooling*, 1 January 2009, REECO Landesmesse Stuttgart, Germany.

[34] Bureau of Meteorology (BOM) Australia. www.bom.gov.au/climate/map/heating-cooling-degree-days/documentation.shtml. Last accessed 13 May 2013.

[35] Danfoss Turbocor chiller product specification sheet. www.turbocor.com/uploaded/Specs_TT500.pdf. Last accessed 30 October 2013.

Index

Note: Boxes, Figures and Tables are indicated by italic page numbers, glossary [and key terms at chapter/section beginnings] by bold numbers